Annals of Mathematics Studies

Number 105

MULTIPLE INTEGRALS IN THE CALCULUS OF VARIATIONS AND NONLINEAR ELLIPTIC SYSTEMS

BY

MARIANO GIAQUINTA

PRINCETON UNIVERSITY PRESS

PRINCETON, NEW JERSEY

1983

Clothbound editions of Princeton University Press books are printed on acid-free paper, and binding materials are chosen for strength and durability. Paperbacks, while satisfactory for personal collections, are not usually suitable for library rebinding.

ISBN 0-691-08330-4 (cloth)
ISBN 0-691-08331-2 (paper)

Printed in the United States of America
by Princeton University Press, 41 William Street
Princeton, New Jersey

☆

Library of Congress Cataloging in Publication data will
be found on the last printed page of this book

CONTENTS

PREFACE

These notes are an enlarged version of the lectures I gave at the Mathematisches Institut der Universität Bonn during the winter semester of the academic year 1980-81.

The first eight chapters essentially reproduce with slight modifications the text of a first draft which appeared in the Vorlesungsreihe des SFB 72 Bonn in February 1981. The last chapter describes some of the new contributions appeared since then.

In preparing these notes I have taken into account lectures and seminars I have given in these last few years; but mainly I have taken advantage of many discussions I have had with colleagues and friends, among them G. Anzellotti, S. Campanato, J. Frehse, W. Jäger, O. John, J. Nečas, J. Stará and particularly my friends E. Giusti and G. Modica. It is a pleasure for me to take the opportunity and thank them. Moreover, I want to thank S. Hildebrandt for having invited me to give these lectures at the University of Bonn, and for comments and stimulating conversations.

I also would like to acknowledge with gratitude the hospitality of the University of Bonn and the support of SFB 72.

MARIANO GIAQUINTA

Multiple Integrals
in the Calculus of Variations
and Nonlinear Elliptic Systems

MULTIPLE INTEGRALS IN THE CALCULUS OF VARIATIONS AND NONLINEAR ELLIPTIC SYSTEMS

Mariano Giaquinta

0. *Introduction*

The aim of these lectures is to discuss the existence and differentiability of minimum points (or, more generally, of stationary points) of regular functionals in the Calculus of Variations; i.e. functionals of the type

$$\int_{\Omega} F(x, u(x), \nabla u(x))\, dx \tag{0.1}$$

where Ω is an open set in R^n, $n \geq 2$, $u(x) = (u^1(x), \cdots, u^N(x))$, $N \geq 1$, is a (vector valued) function defined in Ω and ∇u stands for $\{D_\alpha u^i\}\alpha = 1, \cdots, n$, $i = 1, \cdots, N$. 'Regular' means that $F(x, u, p)$ is convex in p.

In 1900 D. Hilbert posed the following two problems in his well-known lecture delivered before the International Congress of Mathematicians at Paris $(n=2, N=1)$:

1. (20th problem) "Has not every regular variation problem a solution, provided certain assumptions regarding the given boundary conditions are satisfied, and provided also if need be that the notion of a solution shall be suitably extended?"

2. (19th problem) "Are the solutions of regular problems in the Calculus of Variations always necessarily analytic?"

These two problems have originated such a great deal of work that it would be very difficult even to quote the different contributions. For an account of some of them we refer to the conferences of E. Bombieri [31] and J. Serrin [269] and to the nowaday classical books of O. A. Ladyzhenskaya and N. N. Ural'tseva [191] and C. B. Morrey, Jr. [231] (for the case $N=1$).

Starting with the remarkable result of S. Bernstein in 1904 that any C^3 solution of a single elliptic nonlinear analytic equation in two variables is necessarily analytic, and through the work of many authors (among others L. Liechtenstein, E. Hopf, T. Radò, J. Schauder, J. Leray, R. Caccioppoli, I. Petrowsky, C. B. Morrey, Jr., etc.) it was proved that every sufficiently smooth stationary point of (0.1), say C^1, is analytic, provided F is analytic.

On the other hand, by direct methods in general one can only prove the existence of weak stationary points in Sobolev spaces.

So the following problem arises: *are weak minimum points of class* C^1 ?

This gap from Sobolev spaces up to C^1 was filled in 1957 by the fundamental work of E. De Giorgi [69] in the scalar case $N = 1$.

All the attempts to fill the same gap in the vector valued case $N > 1$ were unfruitful, and in 1968 E. De Giorgi [71] and E. Giusti and M. Miranda [138] gave examples of functionals of the type (0.1) with noncontinuous minimum points.

The works of C. B. Morrey, Jr. [232] and E. Giusti and M. Miranda [139] then start the study of the 'partial regularity' in the vector valued case. And in these notes we shall confine ourselves to report on some recent results on the partial (which means regularity except on a closed 'small' singular set) and everywhere regularity, especially in the vector valued case $N > 1$. Some backgrounds are also presented.

Our main goal will be to point out some of the methods that have been introduced. Many important topics, such as existence and uniqueness, have only been hinted at. So we are far from being complete. Moreover applications such as to differential geometry or to problems in mechanics are omitted.

For the plan of the notes we refer to the contents. We only mention that as far as possible, each chapter and each section are written indepen dently of the others, even despite of shortness.

Chapter I

MULTIPLE INTEGRALS IN THE CALCULUS OF VARIATIONS: SEMICONTINUITY, EXISTENCE AND DIFFERENTIABILITY

1. *Multiple integrals: first and second variation*

Let Ω be an open set in the Euclidean n-dimensional space R^n. For the sake of simplicity let us assume that Ω be a bounded and connected open set with smooth boundary; moreover, assume that $n \geq 2$. We shall denote by $x = (x_1, \cdots, x_n)$ points in R^n and by $dx = dx_1, \cdots, dx_n$ or $d\mathcal{L}^n$ the Lebesgue volume element in R^n.[1)]

Let $u(x) = (u^1(x), \cdots, u^N(x))$ be a vector valued function defined in Ω with value in R^N, $N \geq 1$. We shall denote by Du or ∇u the gradient of u, i.e. the set $\{D_\alpha u^h\}$ $\alpha = 1, \cdots, n$; $h = 1, \cdots, N$, where $D_\alpha = \frac{\partial}{\partial x_\alpha}$.

We shall consider multiple integrals of the form

$$J[u] = \int_\Omega F(x, u(x), \nabla u(x))\, dx \tag{1.1}$$

where $F(x, u, p): \Omega \times R^N \times R^{nN} \to R$. Of course the domain of the functional J will be a class of functions, which we shall call *admissible functions*. We shall look at the problem of *minimizing the functional* $J[u]$ *among the admissible functions*. Then we want *to study qualitative properties of these minimum points* (assuming that there exists at least one) or more generally of the *stationary points*, which are called in this setting *extremals of* J.

1) We shall also use the notation 'meas A' or '$|A|$' for $\mathcal{L}^n(A)$.

Simple examples of functionals of type (1.1) are the *Dirichlet integral*:

$$D[u] = \int_{\Omega} |\nabla u|^2 dx$$

and the *area integral*:

$$A[u] = \int_{\Omega} \sqrt{1 + |\nabla u|^2}\, dx \, .$$

Both these two functionals are defined on scalar functions, i.e. $N = 1$; variational integrals defined on classes of vector valued functions appear for instance in the mathematical theory of linear and nonlinear elasticity, in the theory of plasticity or elastoplasticity and in differential geometry, for example when studying H-surfaces or minimal immersions.

It is worth remarking that the relations between minimum problems for functionals (1.1) and boundary value problems for a class of partial differential systems (elliptic systems) are very strict, as we shall see.

In this section we start by recalling a few classical results from the Calculus of Variations, such as Euler equations and second variation for the functional J in (1.1).

Let us assume that the function J in (1.1) is of class C^1, and let us consider the functional $J[u]$ as defined in $C^1(\overline{\Omega}, R^N)$. For a given $\phi \in C^1(R^n, R^N)$, let

$$K = \{u \in C^1(\overline{\Omega}, R^N) : u = \phi \text{ on } \partial\Omega\}$$

be the class of admissible functions; and assume that u be a minimum point for J in K. For $v \in C_0^{\infty}(\Omega, R^N)$ and $t \in R$, the function $u + tv$ is still an admissible function (or as it is usually said tv is an *admissible variation*); therefore we must have

$$\frac{d}{dt} J[u+tv]_{|t=0} = 0$$

i.e.

$$(1.2) \qquad \int_\Omega [F_{p_\alpha^i}(x,u,\nabla u) D_\alpha v^i + F_{u^i}(x,u,\nabla u) v^i] dx = 0 \qquad \forall v \in C_0^\infty(\Omega, R^N) .$$

Here and in the following we use the convention that repeated indices are summed: here α goes from 1 to n and i from 1 to N.

We shall call the left-hand side of (1.2), as in current literature, the *first variation of* J *at* u and (1.2) the *Euler equation* (*in its weak formulation*) *of* J *at* u.

If moreover we assume J, Ω, u sufficiently smooth (for example of class C^2) then we can integrate by parts in (1.2) getting

$$\int_\Omega [-D_\alpha F_{p_\alpha^i}(x,u,\nabla u) + F_{u^i}(x,u,\nabla u)] v^i dx = 0 \qquad \forall v \in C_0^\infty(\Omega, R^N)$$

and hence

$$(1.3) \qquad -D_\alpha F_{p_\alpha^i}(x,u,\nabla u) + F_{u^i}(x,u,\nabla u) = 0 \quad \text{in} \quad \Omega \qquad i = 1, \cdots, N$$

i.e.

$$F_{p_\alpha^i p_\beta^j}(x,u,\nabla u) D_\alpha D_\beta u^j + F_{p_\alpha^i u^j}(x,u,\nabla u) D_\alpha u^j + F_{p_\alpha^i x_\alpha}(x,u,\nabla u) -$$

$$- F_{u^i}(x,u,\nabla u) = 0 \qquad i = 1, \cdots, N$$

which is a *quasilinear* system of partial differential equations: linear with respect to the second derivatives and nonlinear with respect to the first and zero order derivatives. (1.3) is called the *Euler equation* (*in its strong formulation*) *of* J *at* u.

For example, the Euler equation for the Dirichlet integral is

$$\Delta u = 0$$

where $\Delta = \sum_{i=1}^{n} \frac{\partial^2}{\partial x_i^2}$ is the Laplace operator, while the Euler equation for the area functional is[2)]

$$\sum_{i=1}^{n} \frac{\partial}{\partial x_i} \frac{D_i u}{\sqrt{1+|\nabla u|^2}} = 0 .$$

REMARK 1.1. In deriving the Euler equation for the functional J we have considered the *Dirichlet problem*, i.e. we chose as admissible functions the functions $u \in C^1(\bar{\Omega}, R^N)$ with prescribed value ϕ on the boundary of Ω. Let us now assume that u minimizes the functional J among all functions $v \in C^1(\bar{\Omega}, R^N)$. Then all functions $v \in C^1(\bar{\Omega}, R^N)$ are admissible variations; hence we get

$$\int_\Omega [F_{p_\alpha^i}(x, u, \nabla u) D_\alpha v^i + F_{u^i}(x, u, \nabla u) v^i] dx = 0 \qquad \forall v \in C^1(\bar{\Omega}, R^N)$$

and integrating by parts

$$(1.4) \quad \int_\Omega [F_{u^i} - D_\alpha F_{p_\alpha^i}] v^i dx + \int_{\partial\Omega} \nu_\alpha F_{p_\alpha^i}(x, u, \nabla u) v^i d\sigma = 0 \quad \forall v \in C^1(\bar{\Omega}, R^N).$$

Here $\nu = (\nu_1, \cdots, \nu_n)$ denotes the unit outward normal to $\partial\Omega$. Since (1.4) holds for all $v \in C_0^\infty(\Omega, R^N)$, we deduce that (1.3) also holds and therefore

---

2)Note that the left-hand side represents n times the mean curvature of the surface $\{(x, u(x))\, x \in \Omega\}$; therefore: a surface of least area has zero mean curvature.

$$\int_{\partial\Omega} \nu_\alpha F_{p_\alpha^i} v^i d\sigma = 0 \qquad \forall v \in C^1(\bar{\Omega}, R^N)$$

that is we have the so-called *natural conditions*

$$\nu_\alpha F_{p_\alpha^i}(x, u, \nabla u) = 0 \quad \text{on} \quad \partial\Omega, \quad i = 1, \cdots, N .$$

For instance if u minimizes the Dirichlet integral without boundary conditions, then it is a solution of the *Neumann problem*:

$$\begin{cases} \Delta u = 0 & \text{in } \Omega \\ \frac{du}{d\nu} = \nu_\alpha D_\alpha u = 0 & \text{on } \partial\Omega . \end{cases}$$

REMARK 1.2. Let us consider the problem

$$\min_{u \in K} J[u]$$

where now K is any convex subset of, say, $C^1(\bar{\Omega}, R^N)$; for example

$$K = \{u \in C^1(\bar{\Omega}, R^N) : u = \phi \quad \text{on} \quad \partial\Omega, \ u^i \geq \chi^i \quad \text{in} \quad \bar{\Omega} \quad i = 1, \cdots, N\}$$

(of course we assume $\chi^i \leq \phi^i$ on $\partial\Omega$ in order to grant that $K \neq \emptyset$). Let u be a minimum point and let $v \in K$; for all $t \in [0,1]$

$$tv + (1-t)u = u + t(v-u)$$

is an admissible function, hence

$$J[u] \leq J[u + t(v-u)] \qquad \forall t \in [0,1]$$

but this time we can only state that

$$\frac{d}{dt} J[u + t(v-u)]_{|t=0} \geq 0$$

i.e.

$$(1.5)\qquad \int_\Omega [F_{p^i_\alpha}(x,u,\nabla u)\,D_\alpha(u^i-v^i)+F_{u^i}(x,u,\nabla u)\,(u^i-v^i)]\,dx \le 0 \qquad \forall v\in K\,.$$

(1.5) is called a *variational inequality.*[3)]

Let us now assume that J be of class C^2 and that $u\in C^1(\overline{\Omega},R^N)$ minimize J among all functions $v\in C^1(\overline{\Omega},R^N)$ with, say, prescribed boundary value. Then for all $\phi\in C^1_0(\Omega,R^N)$ we must have

$$(1.6)\qquad \frac{d^2}{dt^2}J[u+t\phi]_{|t=0}\ge 0\,.$$

If $n=N=1$ and $\Omega=(a,b)$, then (1.6) becomes

$$(1.7)\qquad \int_a^b [F_{pp}(x,u,\dot u)\,\dot\phi^2+2F_{pu}(x,u,\dot u)\,\phi\dot\phi+F_{uu}(x,u,\dot u)\,\phi^2]\,dx\ge 0 \qquad \forall\phi\in C^1_0(a,b)\,.$$

By approximation it follows that (1.7) holds for all Lipschitz functions ϕ which vanish on the boundary. Choosing now, for $x_0\in(a,b)$, ε sufficiently small, $\phi=\phi_\varepsilon$ where

$$\phi_\varepsilon(x)=\begin{cases}0 & \text{if } 0\le x\le x_0-\varepsilon\\ \frac{1}{\varepsilon}(x-x_0)+1 & \text{if } x_0-\varepsilon\le x\le x_0\\ -\frac{1}{\varepsilon}(x-x_0)+1 & \text{if } x_0\le x\le x_0+\varepsilon\\ 0 & \text{if } x_0+\varepsilon\le x\le b\end{cases}$$

and letting ε go to zero, we obtain, for all $x_0\in(a,b)$, the necessary condition

---

3) For more information on variational inequalities see e.g. D. Kinderlehrer, G. Stampacchia [185].

$$F_{pp}(x_0, u(x_0), \dot{u}(x_0)) \geq 0 ,$$

that is called *Legendre-condition.*

If we repeat this derivation for the case $n > 1$, $N = 1$ we obtain for the minimizing point u at each $x_0 \in \Omega$

$$F_{p_\alpha p_\beta}(x_0, u(x_0), \nabla u(x_0))\xi_\alpha \xi_\beta \geq 0 \qquad \forall \xi \in R^n \tag{1.8}$$

which is still called *Legendre-condition.* In the general case $n \geq 2$, $N > 1$ we get instead

$$F_{p^i_\alpha p^j_\beta}(x_0, u(x_0), \nabla u(x_0))\xi_\alpha \xi_\beta \lambda^i \lambda^j \geq 0 \qquad \forall \xi \in R^n, \forall \lambda \in R^N \tag{1.9}$$

which is called *Legendre-Hadamard condition.*

Partially justified by the above considerations is the following definition: *the functional* $J[u]$ *is a regular multiple integral of the Calculus of Variations if the Legendre-Hadamard condition holds with strict inequality for all* $\xi \in R^n - \{0\}$, $\lambda \in R^N - \{0\}$ *and all* (x, u, p).

Let us note that for $N = 1$ the regularity condition is equivalent to the strict convexity of $F(x, u, p)$ with respect to p and to the *ellipticity condition* of the Euler equation, while in the case of systems, $N > 1$, the situation is quite different. The convexity condition (or strong ellipticity) of F with respect to p is

$$F_{p^i_\alpha p^j_\beta}(x, u, \nabla u)\xi^i_\alpha \xi^j_\beta > 0 \qquad \forall \{\xi^i_\alpha\} \neq 0 \tag{1.10}$$

and this condition is stronger than the regularity condition. But in the sequel we shall often speak of regular integrals meaning that (1.10) holds instead of (1.9).

Now one could ask whether the regularity condition is also sufficient in order to have u be a minimum point. The answer is negative in general, and there are classical and elegant theories (for example Weierstrass theory, Hilbert's invariant integral theory, Jacobi field theory) dealing with the problem of finding sufficient conditions for u to be a minimum point; but we shall omit them and refer to the books quoted below. We would only like to note that for functionals of the type

$$J[u] = \int_\Omega F(x, \nabla u)\, dx$$

with $F(x, p)$ convex with respect to p, stationary points are also minimum points (among functions with the same boundary value). In fact from the convexity we have

$$J[u] - J[U] \le \int_\Omega F_{p^i_\alpha}(x, \nabla u)(D_\alpha u^i - D_\alpha U^i)\, dx$$

which is equal to zero since $u = U$ on $\partial\Omega$ and u is an extremal.

In the sequel it will happen that we consider more general variational integrals of the type

$$\int_\Omega F(x, \delta u, D^m u)\, dx$$

where δu stands for $\{D^\alpha u^h\}\, |\alpha| \le m-1$, $h = 1, \cdots, N$, $D^m u$ stands for $\{D^\alpha u^h\}\, |\alpha| = h = 1, \cdots, N, m$ is an integer ≥ 1 and α is a multi-index $\alpha = (\alpha_1, \cdots, \alpha_n)$ (α_i nonnegative integers). Here

$$D^\alpha u = \frac{\partial^{|\alpha|}}{\partial x_1^{\alpha_1}\, \partial x_2^{\alpha_2} \cdots \partial x_n^{\alpha_n}}$$

and $|\alpha| = \alpha_1 + \alpha_2 + \cdots + \alpha_n$.

Let $F = F(x, \eta, \xi)$ and denote by p the variable (η, ξ); the Euler equation is

$$\sum_{h=1}^{N} \sum_{|\alpha|\le m} \int_{\Omega} F_{p_\alpha^h}(x, \delta u, D^m u) D^\alpha \phi^h dx = 0 \qquad \forall \phi \in C_0^\infty(\Omega, R^N)$$

or, in its strong formulation,

$$\sum_{|\alpha|\le m} (-1)^{|\alpha|} D^\alpha F_{p_\alpha^h}(x, \delta u, D^m u) = 0 \quad \text{in } \Omega \quad h = 1, \cdots, N$$

which is a system of quasilinear partial differential equations of order $2m$.

In this case the Legendre-Hadamard condition becomes

$$\sum_{|\alpha|=|\beta|=m} F_{p_\alpha p_\beta}(x, \delta u, D^m u) \xi^{\alpha+\beta} \ge 0 \qquad \forall \xi \in R^n$$

if $n \ge 2$ and $N = 1$, $\xi^\alpha = \xi_1^{\alpha_1} \cdot \xi_2^{\alpha_2} \cdot \ldots \cdot \xi_n^{\alpha_n}$, while it assumes the form

$$\sum_{ij=1}^{N} \sum_{|\alpha|=|\beta|=m} F_{p_\alpha^i p_\beta^j}(x, \delta u, D^m u) \xi^{\alpha+\beta} \lambda_i \lambda_j \ge 0 \qquad \forall \xi \in R^n, \forall \lambda \in R^N$$

if $n \ge 2$, $N > 1$.

It may happen that we shall also consider functionals for which the order of derivatives of u depends on which component we are considering. These functionals give rise to Euler equations of the following type

$$\sum_{i=1}^{N} \sum_{|\alpha|\le m_i} \int_{\Omega} A_i^\alpha(x, Du) D^\alpha \phi^i dx = 0 \qquad \forall \phi \in C_0^\infty(\Omega, R^N)$$

where this time Du stands for $\{D^\alpha u^i\}$ $i = 1, \cdots, N$, $|\alpha| \le m_i$ and m_i are nonnegative integers.

Finally, for more information, particularly on the classical theory, one may refer to any of the many wonderful books on the Calculus of Variations, such as the ones by N. T. Akhiezer, G. A. Bliss, O. Bolza, C. Caratheodory, R. Courant, P. Funk, J. M. Gelfand and V. Fomin, J. Hadamard, M. Morse, G. Talenti, L. Tonelli, L. C. Young and others; in particular one may refer to C. B. Morrey [231] and S. Hildebrandt [156] for what concerns our treatment, and for the one-dimensional case to the lecture notes by S. Hildebrandt [157], where one can also find historical references.

2. *Semicontinuity theorems*

Let us first recall a few general topological facts. Let X be a topological space. We say that $J : X \to \mathbf{R} \cup \{+\infty\}$ is *lower semicontinuous* (l.s.c.) if for all $a \in \mathbf{R}$

$$U_a^J = \{x \in X : J(x) > a\}$$

is an open subset of X, or equivalently

$$V_a^J = \{x \in X : J(x) \leq a\}$$

is a closed subset of X. It is not difficult to prove the following characterization of l.s.c. functions: $J : X \to \mathbf{R} \cup \{+\infty\}$ *is* l.s.c. *if and only if*

$$E^J = \{(x,a) \in X \times \mathbf{R} : J(x) \leq a\}$$

is a closed subset in $X \times \mathbf{R}$.

The following proposition will be more useful for us:

PROPOSITION 2.1. *If* $J : X \to \mathbf{R} \cup \{+\infty\}$ *is l.s.c., then for* $x = \lim_{i\to\infty} x_i$, *we have*

$$J(x) \leq \liminf_{i\to\infty} J(x_i) .$$

If moreover X *satisfies the first countability axiom (i.e. for all* $x \in X$, *there exists a countable fundamental system of neighborhoods), then the converse holds.*

We have moreover: *let* $\{J_i\}_{i\in I}$ *be a family of l.s.c. functions, then* $J = \sup_i J_i$ *is l.s.c.*

The next theorem plays a fundamental role in the following

THEOREM 2.2 (Weierstrass). *Let* $J : X \to R \cup \{+\infty\}$ *be a l.s.c. function and let* K *be a subset of* X *satisfying one of the two following conditions*:

i) K *is a compact subset of* X.

ii) K *is a sequentially compact subset of* X (*i.e. from every sequence of points in* K *we can select a subsequence converging to a point of* K).

Then there exists a minimum point for J *in* K.

We can state a little bit more. The function J will be called *sequentially lower semicontinuous (s.l.s.c.)* if for $x_i \to x$ in X we have

$$J(x) \leq \liminf_{i\to\infty} J(x_i) .$$

Then we have

THEOREM 2.2′. *Let* $J : X \to R \cup \{+\infty\}$ *be s.l.s.c. and let* K *be a sequentially compact subset of* X. *Then there exists a minimum point for* J *in* K.

We note that if X satisfies the first countability axiom, then compact subsets of X are sequentially compact; and if X is a metric space, then compactness and sequential compactness are equivalent. Finally we recall that the supremum of s.l.s.c. functions is a s.l.s.c. function.

Let us now consider the problem of minimizing a function J defined on a set K. Generally, K is not equipped a priori with a topology. So our minimum problem can be seen as the problem of introducing a topology on K for which K is a sequentially compact set and J is a s.l.s.c. function. Note that in order to grant that J be s.l.s.c. we need in general a rich topology, while for the compactness of K the topology need not be too rich.

We shall see that this compromise can be satisfactorily reached for a large class of multiple integral functionals working on Sobolev spaces $H^{k,m}(\Omega, R^N)$.

We recall that the space $H^{k,m}(\Omega)$ is the space of the functions $u \in L^m(\Omega)$ with (distributional) derivatives up to order k in $L^m(\Omega)$.

Let Ω be a bounded, connected open subset of R^n and let F *be a function*

$$F(x,y,z): \Omega \times R^N \times R^m \to R$$

such that

i) $F \geq 0$

ii) F *is measurable in* x *for all* (y,z)

iii) F *is continuous in* y *for all* z *and almost every* x

iv) F *is convex in* z *for all* y *and almost every* x.

Then, for almost every x, $F(x,y,z)$ is a continuous function with respect to (y,z),[4)] therefore it follows that if

$$u : \Omega \subset R^n \to R^N \qquad u(x) = (u^1(x), \cdots, u^N(x))$$
$$p : \Omega \subset R^n \to R^m \qquad p(x) = (p^1(x), \cdots, p^m(x))$$

are measurable functions, then $x \to F(x, u(x), p(x))$ is measurable in Ω.[5)] Hence we can consider the functional

$$J[u,p] = \int_\Omega F(x, u(x), p(x))\, dx$$

which may also have value $+\infty$.

[4)]PROPOSITION. *Let* $f(y,z)$ *be continuous with respect to* $y \in R^k$ *for all* $z \in R^m$ *and convex with respect to* z *for all* y. *Then*

i) $z \to f(y,z)$ *are equicontinuous functions for* y *lying on a compact subset of* R^k

ii) f *is continuous in the product space* $R^k \times R^m$.

[5)]PROPOSITION. *Let* $h(x,y)$ *be measurable with respect to* $x \in R^n$ *for all* $y \in R^k$ *and continuous in* y *for almost every* x. *Let* $w : R^n \to R^k$ *be a measurable function. Then* $x \to h(x, w(x))$ *is measurable.*

In the applications it will often be $m = n \cdot N$ and $p(x) = \nabla u(x)$, but it is more convenient to have results in this more general situation.

We now want to prove

THEOREM 2.3. *Let* Ω *be a bounded open set in* R^n *and let* $F(x,y,z): \Omega \times R^N \times R^m \to R$ *be a function satisfying i) ii) iii) iv). Suppose that* $\{u_h\}$ *converges to* u *in* $L^1(\tilde{\Omega}, R^N)$ *and* $\{p_h\}$ *converges weakly to* p *in* $L^1(\tilde{\Omega}, R^m)$ *for each* $\tilde{\Omega} \subset\subset \Omega$.[6] *Then*

$$J[u,p] \le \liminf_{h\to\infty} J[u_h, p_h]$$

i.e. J *is s.l.s.c. in* $L^1_{loc}(\Omega, R^N) \times L^1_{loc}(\Omega, R^m)_{weak}$.

Proof. We shall divide the proof in two steps.

1st step. Assume, besides the hypotheses of Theorem 2.3, that *there exist the derivatives of* F *with respect to* z *and that* (v) $F_z(x,y,z)$ *is measurable in* x, *continuous in* (y,z) *for a.e.x, then the conclusion of Theorem 2.3 holds.*

Let us choose a sequence, which we still call (u_h, p_h) such that

$$J[u_h, p_h] \to \liminf_{h\to\infty} J[u_h, p_h] .$$

We can also assume that $u_h(x) \to u(x)$ a.e. in $\tilde{\Omega} \subset\subset \Omega$. For a fixed $\Omega' \subset\subset \Omega$, suppose that

$$J_{\Omega'}[u,p] = \int_{\Omega'} F(x,u,p)\,dx < +\infty .$$

[6] $\tilde{\Omega} \subset\subset \Omega$ means that the closure of $\tilde{\Omega}$ is contained in Ω, i.e. $\bar{\tilde{\Omega}} \subset \Omega$.

From Egorov's[7] and Lusin's[8] theorems, we deduce that for all $\varepsilon > 0$ there exists a compact subset $K \subset \Omega'$ such that u and p are continuous on K, $u_h \to u$ uniformly on K and that

$$\int_{\Omega' \setminus K} F(x,u,p)\,dx < \varepsilon \,.^{9)} \tag{2.1}$$

(If $J_{\Omega'}[u,p] = +\infty$, we can find $K \subset \Omega'$ such that $J_K[u,p] > \frac{1}{\varepsilon}$.) Since F is convex in z, we obtain

$$\begin{aligned} F(x,u_h,p_h) &\geq F(x,u_h,p) + \sum_{i=1}^{m} \frac{\partial F}{\partial z_i}(x,u_h,p)(p_h^i - p^i) = \\ &= F(x,u_h,p) + \sum_{i=1}^{m} \frac{\partial F}{\partial z_i}(x,u,p)(p_h^i - p^i) + \\ &+ \sum_{i=1}^{m} \left[\frac{\partial F}{\partial z_i}(x,u_h,p) - \frac{\partial F}{\partial z_i}(x,u,p)\right](p_h^i - p_h)\,. \end{aligned} \tag{2.2}$$

Suppose now that *the vector* $F_z(x,y,z)$ *be continuous in* (x,y,z): then, since $p_h \rightharpoonup p$ [10] in $L^1(\Omega', R^m)$ and $F_z(x,u,p)$ is bounded on K, we get

---

[7] *Egorov's theorem: Let* f_h, $f: A \to Y$, *where* A *is a* μ*-measurable set with* $\mu(A) < +\infty$ *and* Y *is a separable space. Suppose that* $f_h \to f$ μ-a.e. *Then for all* $\varepsilon > 0$ *there exists a measurable subset* B *with* $\mu(A \setminus B) < \varepsilon$ *such that* $f_h \to f$ *uniformly on* B.

[8] *Lusin's theorem: Let* μ *be a Borel (or Radon) measure; let* f *be a* μ*-measurable function with value on a separable metric space* Y *and let* A *be a* μ*-measurable subset with* $\mu(A) < +\infty$. *Then for all* $\varepsilon > 0$ *there exists a closed (compact) subset* C *such that* $\mu(A \setminus C) < \varepsilon$ *and* $f_{|C}$ *is continuous.*

[9] We use the *Absolute Continuity theorem: Let* f *be* μ*-summable. Then for all* $\varepsilon > 0$ *there exists* $\delta > 0$ *such that* $\int_A |f|\,dx < \varepsilon$ *for all measurable set* A *with* $\mu(A) < \delta$.

[10] By $\rightharpoonup$ we mean weak convergence.

$$\lim_{h\to\infty} \int_K \sum_{i=1}^m \frac{\partial F}{\partial z_i}(x, u, p)(p_h^i - p^i)\, dx = 0 .$$

On the other hand, because of the Banach-Steinhaus theorem, the L^1-norms of p_h and p are equi-bounded, $F_z(x, u_h, p)$ converges uniformly to $F_z(x, u, p)$ in K, hence

$$\lim_{h\to\infty} \int_K \left[\sum_{i=1}^m \frac{\partial F}{\partial z_i}(x, u_h, p) - \frac{\partial F}{\partial z_i}(x, u, p)\right] (p_h^i - p^i)\, dx = 0 .$$

Therefore, taking into account (2.1) and (2.2), it follows

$$J_{\Omega'}[u, p] - \varepsilon \leq J_K[u, p] = \lim_{h\to\infty} \int_K F(x, u_h, p)\, dx \leq$$

$$\leq \liminf_{h\to\infty} \int_K F(x, u_h, p_h)\, dx \leq \liminf_{h\to\infty} J_\Omega[u_h, p_h] .$$

If $J_{\Omega'}[u, p] = +\infty$, we have

$$\frac{1}{\varepsilon} < J_K[u, p] \leq \liminf_{h\to\infty} J[u_h, p_h] .$$

This gives the result in step 1 under the stronger condition that F_z be continuous. Now we want to get rid of such a condition, and this can be done by using the following lemma which we state without proof:

LEMMA 2.1. *Let* Ω *be a* μ*-measurable subset of* R^n *with* $\mu(\Omega) < +\infty$. *Let* $h(x, y) : \Omega \times R^\ell \to R$ *be measurable in* x *and uniformly continuous in* y *for* a.e.x. *Then for each* $\delta > 0$ *there exists a closed subset* $\Omega_\delta \subset \Omega$, *with* $\mu(\Omega \setminus \Omega_\delta) < \delta$, *such that* $h(x, y)$ *is continuous in* $\Omega_\delta \times R^\ell$.

Let us go back to the proof of step 1. Since the L^∞-norms of u, u_h and p are equi-bounded by a constant L on K, the function $F_z(x,y,z)$ satisfies the hypotheses of Lemma 2.1 in

$$K \times \{y \in R^N : |y| \leq L\} \times \{z \in R^m : |z| < L\} = K \times A \times B .$$

Therefore there exists a closed set $\Omega'' \subset K$ (Ω'' compact) such that

$$F_z(x,y,z) \text{ is continuous in } \Omega'' \times A \times B$$

$$J_{\Omega''}[u,p] \geq J_K[u,p] - \varepsilon .$$

And now we can repeat the proof given above in Ω'' and get the result.

2nd step. In order to get rid of the differentiability assumption on F we use the following remark, see [72]. Under the assumption of Theorem 2.3, the function $F(x,y,z)$ can be obtained as the limit of a nondecreasing sequence of nonnegative functions $F_j(x,y,z)$ which are measurable on x, infinitely differentiable in (y,z) for a.e.x. and convex in z for a.e.x. and for all y. Then, from step 1,

$$J^{(j)}_{\Omega'}[u,p] = \int_{\Omega'} F_j(x,u,p)\,dx$$

are s.l.s.c. in $L^1(\Omega', R^N) \times L^1(\Omega, R^m)_{weak}$ for all $\Omega' \subset \Omega$, and from this the conclusion of the theorem follows immediately taking the sup in j (and Ω'). q.e.d.

Let us define

$$\phi_t(x) = \begin{cases} 1 & \text{if} \quad |x| \leq t \\ 0 & \text{if} \quad |x| > t \end{cases} \qquad t \in \mathbf{N}$$

then we have

$$\int_\Omega F(x, u, p)\, dx = \sup_t \int_\Omega F(x, u, p)\, \phi_t\, dx$$

hence: *the conclusion of Theorem 2.3 holds for unbounded open subsets of* R^n, *too.*

By using Hölder inequality (for bounded Ω) and the above remark it follows immediately

THEOREM 2.4. *Let* Ω *be an open set in* R^n *and let i) ii) iii) iv) hold. Suppose that for all* $\tilde{\Omega} \subset\subset \Omega$ (u_h, p_h) *converges to* (u, p) *in* $L^s(\tilde{\Omega}, R^N) \times L^q(\tilde{\Omega}, R^m)_{weak}$, $1 \leq s < \infty$, $1 \leq q < \infty$, *then*

$$J[u, p] \leq \liminf_{h \to \infty} J[u_h, p_h]$$

i.e. $J[u, p]$ *is s.l.s.c. in* $L^s(\Omega, R^N) \times L^q(\Omega, R^m)_{weak}$.

Note that Theorem 2.4 obviously extends to the case in which each component u_i, p_i lies in different L^{s_i} and L^{q_i} spaces.

And now we come to the case of variational regular integrals:

THEOREM 2.5. *Let* Ω *be an open set in* R^n. *Assume that i) ... iv) hold with* $m = nN$. *Then the functional*

$$J[u] = \int_\Omega F(x, u, \nabla u)\, dx$$

is s.l.s.c. with respect to the weak convergence in $H^{1,q}_{loc}(\Omega, R^N)$, $1 \leq q < +\infty$.

Proof. It is sufficient to prove the theorem in the case where Ω is bounded. If now $u_h \rightharpoonup u$ in $H^{1,q}_{loc}(\Omega, R^N)$, from Rellich theorem it follows that $u_h \to u$ in $L^p(\tilde{\Omega}, R^N)$ $\forall\ \tilde{\Omega} \subset\subset \Omega$; hence $(u_h, \nabla u_h)$ converges to $(u, \nabla u)$ in $L^q(\tilde{\Omega}, R^N) \times L^q(\tilde{\Omega}, R^{mN})_{weak}$ and the result follows from Theorem 2.4. q.e.d.

Let us note that in order to apply Rellich theorem we need $\partial\Omega$ smooth (for example continuous): for this reason we stated all theorems with convergence in $\tilde{\Omega} \subset\subset \Omega$.

Finally we obviously have:

THEOREM 2.6. *Let* Ω *be an open set in* R^n *and i)... iv) hold. Then the functional*

$$J[u] = \int_\Omega F(x, \delta u, D^m u)\,dx$$

is s.l.s.c. with respect to the weak convergence in $H^{m,q}_{loc}(\Omega, R^N)$ $1 \leq q < +\infty$.

More on semicontinuity. Here we would like to state without proof a few interesting results on semicontinuity.

Let us still consider the functional

$$J[u, v] = \int_\Omega F(x, u, v)\,dx$$

where $F(x, y, z) : \Omega \times R^k \times R^m \to R$ satisfies i)... iv), and moreover

$$0 \leq F(x, y, z) \leq M(1 + |z|^s) \qquad M > 0 \qquad s \geq 1$$

for a.e.x. and for all (y, z). Then we have, see [72]

THEOREM 2.7. *Let* Ω *be a bounded (or such that* meas $\Omega < +\infty$ *) open set in* R^n. *Then*

i) *for all fixed* $u \in L^p(\Omega, R^k)$ $1 \leq p \leq +\infty$, *the functional*

$$v \to J[u, v]$$

is strongly continuous in $L^q(\Omega, R^m)$ *for all* $q \geq s$.

ii) *the functional*

$$u \to J[u, v]$$

is strongly equi-continuous in $L^p(\Omega, R^k)$ *for* v *in a bounded subset of* $L^q(\Omega, R^m)$ $q \geq s$, $1 \leq p < +\infty$.

Since in a normed space a convex subset is closed if and only if it is weakly closed, and the functional

$$v \in L^q(\Omega, R^m) \to J[u, v]$$

is convex, we deduce that

$$v \to J[u, v]$$

is l.s.c. in $L^q(\Omega, R^m)$ $q \geq s$, for all $u \in L^p(\Omega, R^k)$ $1 \leq p \leq +\infty$. Now from the proposition in the footnote 4) we get

THEOREM 2.8. *Under the assumption of Theorem 2.7, the functional* $J[u, v]$ *is l.s.c. in* $L^p(\Omega, R^k) \times L^q(\Omega, R^m)_{weak}$ *for* $1 \leq p \leq +\infty$ *and* $q \geq s$.

From Theorem 2.8 one could now deduce Theorem 2.3 (see [72] [156]). We would like to remark that the assumptions in Theorem 2.7 cannot be weakened.

The second set of results we want to state refers to functionals of the type

$$J[u] = \int_\Omega F(x, Lu, Mu)\, dx \tag{2.3}$$

see [89] [35], where $F : \Omega \times R^k \times R^m \to R$ still satisfies i) ... iv) and L, M are (not necessarily linear) operators defined on a Banach space X

$$L : X \to L^q(\Omega, R^k)$$

$$M : X \to L^q(\Omega, R^m) .$$

Functionals (2.3) play an important role for example in the linear (and maybe nonlinear) elasticity theory.

From Theorem 2.3 it follows

THEOREM 2.9. *Suppose that* L *and* M *are sequentially continuous from* X *with the weak topology in respectively* $L^q(\Omega, R^k)$ *with the strong topology and in* $L^q(\Omega, R^m)$ *with the weak topology. Then the functional* J *in (2.3) is s.l.s.c. with respect to the weak topology of* X.

If moreover we assume that

$$0 \le F(x,y,z) \le C(1+|z|^s) \qquad c > 0, \quad s \ge 1$$

we have

THEOREM 2.10. *Suppose that* L *be sequentially continuous from* X *with the weak topology in* $L^q(\Omega, R^k)$ *with the strong topology and that* M *be linear and continuous from* X *with the strong topology in* $L^q(\Omega, R^m)$, $q \ge s$, *with the strong topology. Then the functional* $J[u]$ *in (2.3) is s.l.s.c. with respect to the weak topology in* X.

Notes. Several semicontinuity results for variational integrals were obtained by L. Tonelli and C. B. Morrey, Jr.; these results were then simplified and extended by J. Serrin in two well-known papers [264][265]. J. Serrin essentially proved

THEOREM 2.11. *Let* $F(x,u,p): \Omega \times R \times R^n \to R$ *be a nonnegative, smooth function which is convex in* p. *Suppose that* $u_h, u \in H^{1,1}_{loc}(\Omega)$ *and* $u_h \to u$ *in* $L^1_{loc}(\Omega)$, *then*

$$\int_\Omega F(x,u,\nabla u)\,dx \le \liminf_{h\to\infty} \int_\Omega F(x,u_h,\nabla u_h)\,dx .$$

Apart from the smoothness assumption, Theorem 2.11 is more general than Theorem 2.3. In fact the equiboundedness of the $L^1(\tilde{\Omega})$-norms of ∇u_h is not needed. Theorem 2.11 was extended to the vector valued case, $N > 1$, by Morrey [231] Theorem 4.4.1, but as it has been shown by G. Eisen [79] the proof is not correct, and moreover the result is not true: for $N > 1$ it is not possible to avoid the boundedness of the L^1-norms of ∇u_k.

Now the literature on the problem of semicontinuity is very broad, among others see [72] [25] [61] [170] [250] [80] and [67] and its bibliography for the semicontinuous extension of variational functionals.

The proof of Theorem 2.3 comes from [231] Theorem 1.8.2 and [72]; while Theorems 2.7, 2.8 are taken from [72] and Theorems 2.9 and 2.10 from [35] [89].

As far as the convexity condition on $F(x, u, p)$ in p is concerned, we would like to note that it is necessary in the scalar case $N = 1$ (classical proofs of this fact are available, see [21] [206] for proofs under sufficiently weak assumptions), but it is very far from the necessity in the vector valued case, $N > 1$. Natural conditions, in the case $N > 1$, would be the Legendre-Hadamard or the quasi-convexity condition of C. B. Morrey [231] Section 4.4.

While we point out the importance of the problem, especially for the applications [19] [20], we refer to [231] Section 4.4, [215] and [19] [21] [22] [23] for results in this direction.[11)]

3. *An existence theorem*

We now want to apply the semicontinuity results of Section 2 to studying some minimum problems for regular multiple integrals.

Let us start with a simple case. Let $F(x, u, p) : \Omega \times \mathbf{R}^N \times \mathbf{R}^{nN} \to \mathbf{R}$ satisfy i)... iv) in Section 2. Moreover let us suppose that

$$F(x, u, p) \geq \nu |p|^m \qquad m > 1, \quad \nu > 0. \tag{3.1}$$

---

11) See also Chapter IX.

We shall consider the problem of minimizing the functional

$$J[u] = \int_\Omega F(x, u, \nabla u)\, dx$$

among functions with a prescribed value on the boundary of Ω.

First we want to make a few remarks on condition (3.1). We recall that a convex function $f(z)$ defined on R^m in general does not have a minimum point: one needs more information on the behavior at infinity. For instance one of the following conditions would be sufficient to ensure the existence of a minimum point for $f(z)$:

(a) f increases at infinity, i.e. for all z there exists $\rho(z)$ such that if $|\zeta| > \rho(z)$ then $f(\zeta) > f(z)$

(b) $\lim_{|z| \to \infty} f(z) = +\infty$

(c) f is coercive, i.e. there exist $\rho > 0$ and $a > 0$ such that for $|z| > \rho \quad f(z) \geq a|z|$

(d) $\lim_{|z| \to \infty} \frac{f(z)}{|z|} = +\infty$.

Note that a condition of the type (3.1) $\Rightarrow$ (d) $\Rightarrow$ (c) $\Rightarrow$ (b) $\Rightarrow$ (a).

The functional $J[u]$ is s.l.s.c. in $H^{1,m}(\Omega, R^N)$ $m \geq 1$; note that $J[u]$ is still s.l.s.c. if instead of $F \geq 0$ we assume

$$F(x, u, p) \geq \nu |p|^m - \chi(x) \qquad \chi \in L^1(\Omega).$$

Let now ϕ be a function defined on $\partial\Omega$ which is trace of a function $\phi \in H^{1,m}(\Omega, R^N)$ for which $J[\phi] < +\infty$. For example, this can be granted by the estimate

$$F(x, u, p) \leq c(1 + |p|^m). \tag{3.2}$$

Then the answer to our existence problem is positive. More precisely, let $\{u_h\}$ be a *minimizing sequence*, i.e.

$$u_h - \phi \in H_0^{1,m}(\Omega, R^N)$$

$$J[u_h] \underset{h\to\infty}{\to} \inf\{J[u] : u \in H^{1,m}(\Omega, R^N) u - \phi \in H_0^{1,m}(\Omega, R^N)\} < +\infty$$

from (3.1) we get

$$\nu \int_\Omega |\nabla u_h|^m dx \leq J[u_h] \leq \text{const independent of } h .$$

On the other hand, from Poincaré inequality, we obtain

$$\int_\Omega |u_h|^m dx \leq \text{const} \left[\int_\Omega |u_h - \phi|^m dx + \int_\Omega |\phi|^m dx \right] \leq$$

$$\leq \text{const} \int_\Omega |\nabla u_h|^m dx + \text{const} \int_\Omega [|\nabla\phi|^m + |\phi|^m] dx .$$

Therefore

$$\|u_h\|_{H^{1,m}(\Omega,R^N)} \leq \text{const independent of } h .$$

Now, as $H^{1,m}(\Omega, R^N)$ is a reflexive Banach space for $m > 1$, passing eventually to a subsequence, we have $\{u_h\}$ converges weakly to a function $u_0 \in H^{1,m}(\Omega, R^N)$ such that $u - \phi \in H_0^{1,m}$ and, because of the semicontinuity theorem,

$$J[u_0] \leq \liminf_{h\to\infty} J[u_h] .$$

Concluding, we have that u_0 minimizes $J[u]$ in the class

$$\{u \in H^{1,m}(\Omega, R^N) : u - \phi \in H_0^{1,m}(\Omega, R^N)\} .$$

We would like to remark that for $m = 1$ we cannot carry on the above argument, because bounded sets of $H^{1,1}(\Omega, R^N)$ are not weakly compact. This is the case for the *area problem*

$$\int_\Omega \sqrt{1 + |\nabla u|^2}\, dx \to \min$$

$$u = \phi \quad \text{on} \quad \partial\Omega .$$

In this case we can estimate uniformly the $H^{1,1}$ norm of a minimizing sequence but we cannot deduce (in fact it is not true) that any subsequence converges weakly in $H^{1,1}$.

Now we want to state a theorem of existence which is general enough. First let us recall the following proposition

PROPOSITION 3.1. *Let* $\{u_h\} \subset H^{1,1}(\Omega, R^N)$. *Suppose that*

i) $\|u_h\|_{H^{1,1}(\Omega, R^N)} \leq$ const *independent of* h

ii) *the set functions* $\tilde{\Omega} \to \int_{\tilde{\Omega}} |\nabla u_h| dx$, $\tilde{\Omega} \subset \Omega$, $h \in \mathbf{N}$, *are uniformly absolutely continuous, i.e.* $\forall \varepsilon > 0$, *there exists* $\delta(\varepsilon) > 0$ *such that if* meas $\tilde{\Omega} < \delta$, *then*

$$\int_{\tilde{\Omega}} |\nabla u_h| dx < \varepsilon \qquad \forall\, h \in \mathbf{N} .$$

Then there exists a subsequence of $\{u_h\}$ *weakly converging in* $H^{1,1}(\Omega, R^N)$ *to some function* $u \in H^{1,1}(\Omega, R^N)$. *On the other hand, if* $\{u_h\}$ *is weakly converging in* $H^{1,1}(\Omega, R^N)$ *then i) and ii) hold.*

Now assume that

$$F(x, u, p) \geq \nu |p| , \quad \nu > 0$$

then, for a minimizing sequence, i) of Proposition 3.1 holds. The idea is that if one has also

$$|p|^{-1} F(x,u,p) \to +\infty \quad \text{for} \quad |p| \to +\infty$$

then ii) of Proposition 3.1 also holds. This is expressed by the following proposition

PROPOSITION 3.2.[12] *Let* $f_0(z)$ *be a continuous function such that*

$$\lim_{|z|\to\infty} |z|^{-1} f_0(z) = +\infty .$$

Then for each M *there exists a function* $\phi(\rho), \phi(\rho) > 0$, $\lim_{\rho\to 0} \phi(\rho) = 0$, *such that*

$$\int_{\tilde{\Omega}} |p(x)| dx \leq \phi(\text{meas } \tilde{\Omega})$$

for all $\tilde{\Omega}$ *and* $p(x)$, *provided that*

$$\int_{\tilde{\Omega}} f_0(p(x))\, dx \leq M .$$

Now we can state the following theorem

THEOREM 3.1.[13] *Let* Ω *be a bounded open set with smooth boundary. Suppose that*

i) $F(x,u,p): \Omega \times R^N \times R^{nN} \to R$ *be measurable in* x *for all* (u, p), *continuous in* u *for a.e.x. and all* p, *convex in* p *for a.e.x. and all* u

12) See [231] Lemma 1.9.1 for the proof.

13) See [231] Theorem 1.9.1 for the proof.

ii) *there exist a continuous function* $f_0(z) : R^{nN} \to R$ *such that*

$$F(x,u,p) \geq f_0(p) \qquad \forall u, \ \forall \text{a.e.} x$$

$$\lim_{|z|\to\infty} |z|^{-1} f_0(z) = +\infty$$

iii) *(boundary data)* X^* *be a nonempty family of functions* $\Omega \to R^N$ *weakly sequentially compact in* $H^{1,1}(\Omega, R^N)$

iv) X *be a nonempty family of functions* $\Omega \to R^N$ *weakly sequentially closed in* $H^{1,1}(\Omega, R^N)$ *such that each* $u \in X$ *coincides on* $\partial\Omega$ *with a function* $u^* \in X^*$

v) *there exist* $u_0 \in X$ *such that*

$$J[u_0] = \int_\Omega F(x, u_0, \nabla u_0)\,dx < +\infty$$

vi) J *be bounded from below.*

Then the functional

$$J[u] = \int_\Omega F(x,u,\nabla u)\,dx$$

takes on its minimum for some $u \in X$.

Obviously a similar theorem can be stated for functionals of the type

$$\int_\Omega F(x, \delta u, D^m u)\,dx\ .$$

Theorem 3.1 permits the solution of minimum problems of different kinds, such as for example Dirichlet or Neumann type problems, optimal control problems and even problems with (pointwise or integral) constraint. But we shall not insist on the applications.

Interesting existence theorems have been stated in [23] for functionals which satisfy an intermediate condition between the Legendre-Hadamard condition and convexity; we refer to [23] and to its bibliography for such results.[14)]

4. *The direct methods in the Calculus of Variations*

The methods used in Section 3 for proving the existence of a minimum point for the functional $J[u]$ are known as *direct methods in the Calculus of Variations.* As we have seen the idea is to show that:

i) the integral to be minimized is bounded from below (in the class of admissible functions), so that the infimum, and therefore a minimizing sequence, exists

ii) (continuity) the integral to be minimized is s.l.s.c. with respect to some kind of convergence in the class of admissible functions

iii) (compactness) the minimizing sequence (or at least there exists a minimizing sequence which) converges with respect to the convergence in ii) to an admissible function.

Once i), ii) and iii) are established the result follows immediately.

Of course direct methods can be used in class of functions different from Sobolev spaces and for general (not necessarily integral) functionals.

Direct methods were used by Riemann, who obtained many interesting results on the 'geometrical theory of functions' by assuming the 'Dirichlet principle': *there exists a unique function which minimizes the Dirichlet integral*

$$D[u] = \int_\Omega |\nabla u|^2 dx$$

among all functions $u \in C^1(\Omega) \cap C^0(\bar{\Omega})$ *which take on given value on the boundary* $\partial\Omega$; *moreover that function is harmonic on* Ω.

---

14)See also Chapter IX.

Riemann's work was criticized by Weierstrass because boundedness from below does not imply the existence of a minimum point. Probably because of this criticism the direct methods were neglected and it was only[15)] after 1900 that Hilbert [154] and Lebesgue [194] rigorously established in certain important cases the Dirichlet principle by using essentially direct methods.[16)] Later these methods were used and popularized by Tonelli in a series of papers and books. He applied them to many single and double integral problems, working in classes of absolutely continuous functions and with uniform convergence. Tonelli was able to deal only with integral functionals for which

$$F(x,u,p) \geq m|p|^k - K \qquad k > n$$

and with a few particular cases in which $k = n = 2$ (think of Sobolev imbedding theorem).

These difficulties were overcome around 1930 by C. B. Morrey who made use of function classes of the type of the Sobolev ones.

The use of Sobolev spaces simplifies a lot the existence theory for a large class of integral functionals, but it has to be remarked that we pay for this simplification in terms of the regularity problem: are the weak minimum points classical functions?

It has to be noted that direct methods had already been used by Haar, Radò in the class of Lipschitz functions; in particular Haar [146] (see T. Radò [256]) was able to show the existence of a unique solution in $C^{0,1}(\Omega)$ of the variational problem

$$\begin{cases} \displaystyle\int_\Omega F(\nabla u)\,dx \to \min \\ u = \phi \quad \text{on} \quad \partial\Omega \end{cases}$$

---

15)Let us quote the attempt of C. Arzela' [15].

16)See also D. Hilbert [155], B. Levi [201], G. Fubini [100], J. Hadamard [147] among others.

assuming that $n = 2$, $N = 1$, $F(p)$ be strictly convex, Ω be strictly convex and ϕ satisfy the 'three points condition'. These methods were extended to the n-dimensional case by many authors (M. Miranda, G. Stampacchia, P. Hartman, D. Gilbarg, among others).

Analogous methods were also developed by S. Bernstein mainly in the spirit of studying second order partial differential equations (Euler equations) and exploited deeply by J. Serrin, N.S. Trudinger, Bakel'man, etc. (see [17] [18] [129] [268]).

We must also recall that direct methods have been used for the (parametric) Plateau problem by J. Douglas, R. Courant, E. J. McShane, M. Morse, C.B. Tompkins, C.B. Morrey among others and, more recently, by many authors of the German school; one may refer for example to Morrey [231] Chapters 9 and 10, R. Courant [65] and J.C.C. Nitsche [249].

Finally, starting from the pioneering work by E. De Giorgi, E.R. Reifenberg, H. Federer, W.H. Fleming, F.J. Almgren on (geometric measure theory and) Plateau problem and parametric elliptic integrands, see e.g. [4] [64] [85] [134] [10] [219] [32] direct methods were used for the nonparametric area (mean curvature, capillarity...) problem in the class of BV functions (i.e. L^1-functions whose derivatives are Radon measures with bounded total variation) by many authors, among others E. Giusti, M. Miranda, M. Emmer, L. Pepe, U. Massari, L. Simon, K. Gerhardt, M. Giaquinta etc.; one may refer to [133] [136] [219] [10]. Results for general functionals with linear growth

$$J[u] = \int_\Omega F(x, u, \nabla u)\, dx$$

$$|p| \le F(x, u, p) \le c(1 + |p|)$$

have been obtained in [124] [6], see also [8] for the vector valued case (in connection with a problem in elasto-plasticity).

5. *On the differentiability of regular integrals: weak solutions to the Euler equations*

Let us consider the regular integral

$$J[u] = \int_\Omega F(x, u, \nabla u)\, dx \ .$$

Besides the assumptions in Sections 2 and 3, we suppose now that

$$\text{(5.1)} \qquad \left|\frac{\partial F}{\partial p}\right| , \left|\frac{\partial F}{\partial u}\right| \leq a_1 |u|^m + a_2 |p|^m + \gamma(x)$$

$$\gamma \in L^1(\Omega) , \ a_1, a_2 \ \text{nonnegative constants} .$$

Let $\overline{u} \in H^{1,m}(\Omega, R^N)$ be a minimum point in some subset K of $H^{1,m}(\Omega, R^N)$ of admissible functions and let us assume that $\overline{u} + \lambda v$ also belongs to K for all $v \in C_0^\infty(\Omega, R^N)$ and $\lambda \in (-1,1)$. We can consider the differential quotient at zero of the function

$$J(\lambda) = J[\overline{u} + \lambda v]$$

given by

$$\frac{J(\lambda) - J(0)}{\lambda} = \int_\Omega dx \int_0^1 [F_{u^i}(x, \overline{u} + t\lambda v, \nabla\overline{u} + t\lambda\nabla v)\, v^i +$$

$$+ F_{p^i_\alpha}(x, \overline{u} + t\lambda\nabla v, \nabla\overline{u} + \lambda t\nabla v)\, D_\alpha \cdot v^i]\, dt \ .$$

Since

$$|\overline{u} + tv| \leq |\overline{u}| + M$$

$$|\nabla\overline{u} + t\lambda\nabla v| \leq |\nabla\overline{u}| + M$$

and hence

$$\left|\int_0^1 [F_{u^i}(\cdots)v^i + F_{p^i_\alpha}(\cdots)D_\alpha \cdot v^i]dt\right| \le$$

$$\le a_1(|\bar{u}|+M)^m + a_2(|\nabla\bar{u}|+M)^m + \gamma(x)$$

using the uniform summability theorem of Lebesgue[17)] we can pass to the limit for $\lambda \to 0$ under the integral, getting that $\bar{u}$ satisfies

$$\int_\Omega [F_{p^i_\alpha}(x,u,\nabla u)D_\alpha v^i + F_{u^i}(x,\bar{u},\nabla\bar{u})vi]dx = 0 \quad \forall v \in C_0^\infty(\Omega,R^N)\,. \tag{5.2}$$

Note that if Ω is smooth and $1 < m \le n$, since

$$\|u\|_{L^{m^*}(\Omega,R^N)} \le c\|u\|_{H^{1,m}(\Omega,R^N)} \qquad m^* = \frac{mn}{n-m} = \text{Sobolev exponent}$$

we can assume instead of (5.1)

$$|F_p|, |F_u| \le a_1|u|^{m^*} + a_2|p|^m + \gamma(x) \tag{5.1$'$}$$

while if $m > n$ instead of m^* in (5.1)′ we can take any exponent $1 \le r < +\infty$.

We could now say that $\bar{u}$ is a weak solution of the Euler equation of the functional J, but this is not fruitful for our next considerations, especially in connection with the regularity theory, see for example [266] [191][231]. It is more convenient to refer as weak solutions to the 'stationary points of J', i.e. functions for which the first differential of J is zero. But in order to differentiate the functional J we need a few more assumptions.

---

17)THEOREM. *Let* $\{f_h\}$ *be a sequence of uniformly summable functions (i.e.* $\forall \varepsilon > 0\ \exists\, \sigma(\varepsilon)$ *such that*

$$E \subset \Omega \quad \text{meas } E < \sigma(\varepsilon) \Rightarrow \int_E |f_h| < \varepsilon\,)$$

converging pointwise. Then

$$\int_\Omega \lim_{h\to\infty} f_h\,dx = \lim_{h\to\infty} \int_\Omega f_h(x)\,dx\,.$$

First let us consider the simple case

$$J[u] = \int_\Omega F(x, u, \nabla u)\,dx$$

with

$$|p|^2 - \lambda(x) \le F(x, u, p) \le c|p|^2 + \chi(x)$$

where λ and χ are nonnegative L^1-functions, and assume moreover that F be of class C^1. At least formally, if J is differentiable in $H^{1,2}$, then its differential at u must be

$$(5.3)\qquad \xi \in H^{1,2}(\Omega, R^N) \to \int_\Omega [F_{p^i_\alpha}(x, u, \nabla u) D_\alpha \xi^i + F_{u^i}(x, u, \nabla u)\xi^i]dx\,.$$

For a moment assume $n \ge 3$. Noting that $\nabla\xi \in L^2(\Omega, R^{nN})$ and (because of Sobolev theorem) $\xi \in L^{2^*}(\Omega, R^N)$, in order to have (5.3) have a meaning we must assume that

$$F_{p^i_\alpha}(x, u, \nabla u) \in L^2(\Omega) \qquad i = 1, \cdots, N,\ \ \alpha = 1, \cdots, n$$

$$F_{u^i}(x, u, \nabla u) \in L^{2^{*\prime}}(\Omega) \qquad i = 1, \cdots, N$$

$$2^{*\prime} = \text{the dual exponent of } 2^* = \frac{2n}{n+2}\,.$$

This is granted, taking into account Sobolev's theorem, for example by the following growth conditions:

$$(5.4)\qquad \begin{cases} |F_p(x, u, \nabla u)| \le \mu[\chi_1(x) + |u|^{\frac{n}{n-2}} + |p|] \\[2ex] |F_u(x, u, \nabla u)| \le \mu[\chi_2(x) + |u|^{\frac{n+2}{n-2}} + |p|^{1+\frac{2}{n}}] \\[2ex] \chi_1 \in L^2(\Omega),\ \ \chi_2 \in L^{\frac{2n}{n+2}}(\Omega) \end{cases}$$

or more simply by

$$|F_u|,\ |F_p| \leq \mu V \qquad V = (1 + |u|^2 + |p|^2)^{\frac{1}{2}} .$$

If $n = 2$, condition (5.4) becomes

$$|F_p(x, u, p)| \leq \mu[\chi_1(x) + |u|^{\frac{q}{2}} + |p|]$$

$$(5.4)' \qquad |F_u(x, u, p)| \leq \mu[\chi_2(x) + |u|^{q-1} + |p|^{2\left(1 - \frac{1}{q}\right)}]$$

$$\chi_1 \in L^2(\Omega),\ \chi_2 \in L^{\frac{q}{q-1}}(\Omega),\ 1 < q < +\infty .$$

Now it is easy to verify that conditions (5.4)(5.4)′ are also sufficient for the differentiability of $J[u]$. Precisely we have:

THEOREM 5.1. *Let* $F(x,u,p) : \Omega \times R^N \times R^{nN} \to R$ *be measurable in* x *and of class* C^1 *in* (u,p) *for a.e.x. Suppose*

$$\nu V^2 - \lambda(x) \leq F(x, u, p) \leq \mu V^2 + \chi(x) \qquad \lambda, \chi \in L^1(\Omega)$$

and that conditions (5.4)′ for $n = 2$ *and (5.4) for* $n \geq 3$ *hold. Then the functional* $J[u]$ *is differentiable in* $H^{1,2}(\Omega, R^N)$ *and its first differential at* u *is given by (5.3).*[18)]

Partially justified by this theorem it is usual to introduce, assuming in addition F of class C^2, the following set of conditions

$$(I) \quad \begin{cases} \nu V^2 - \lambda \leq F(x, u, p) \leq \mu V^2 & \nu, \mu, \lambda > 0 \\ |F_p|, |F_{px}|, |F_u|, |F_{ux}| \leq \mu V & \\ |F_{pu}|, |F_{uu}| \leq \mu & \\ m|\xi|^2 \leq F_{p^i_\alpha p^j_\beta}(x, u, p)\xi^i_\alpha \xi^j_\beta \leq M|\xi|^2 & \forall \xi;\ m, M > 0 \end{cases}$$

---

18) See [231] Theorem 1.10.3 for the proof.

and to refer to them as the 'common conditions of Morrey' or 'the natural assumptions of Ladyzhenskaya and Ural'tseva'. Assumptions (I) reduce simply to

$$
(\mathrm{I})_{\mathrm{bis}} \quad \begin{cases} \nu\tilde{V}^2 - \lambda \le F(x,p) \le \mu\tilde{V}^2 \\ |F_p|, |F_{px}| \le \mu\tilde{V} \\ m|\xi|^2 \le F_{p^i_\alpha p^j_\beta}(x,p)\,\xi^i_\alpha \xi^j_\beta \le M|\xi|^2 \qquad \forall\xi;\ m, M > 0 \\ \tilde{V} = (1+|p|^2)^{1/2} \end{cases}
$$

in case $F = F(x,p)$ does not depend explicitly on u.

Note that (I) can be weakened taking into account Sobolev's theorem as in the beginning of this section.

While assumptions $(\mathrm{I})_{\mathrm{bis}}$ are 'natural', the same is not true for conditions (I) or (I) weakened by using Sobolev's theorem, i.e. (5.4)(5.4)′. In fact it is quite unnatural to assume that F_u increases of the same order, with respect to p, as F_p. For instance, for the simple functional

$$
J[u] = \int_\Omega a(u)|\nabla u|^2\,dx
$$

$$
N = 1,\ 0 < m \le a(u) \le M,\ m \le a'(u) \le M
$$

we have

$$
|F_p| \sim |p|, \qquad F_u = a'(u)|p|^2 \sim |p|^2
$$

and not $|F_u| \sim |p|$ or $|F_u| \sim |p|^{1+\frac{2}{n}}$ $n > 2$, $|F_u| \sim |p|^{2\ 1-\frac{1}{q}}$ $1 < q < +\infty$ for $n = 2$.

Because of that, it is usual to introduce a second set of assumptions still called 'natural assumptions of Ladyzhenskaya and Ural'tseva' or 'common conditions of Morrey' and precisely

$$\text{(II)} \quad \begin{cases} \nu\tilde{V}^2 - \lambda \leq F(x,u,p) \leq \mu(R)\tilde{V}^2 \\ |F_u|, |F_{ux}|, |F_{uu}| \leq \mu(R)\tilde{V}^2 \\ |F_p|, |F_{up}|, |F_{xp}| \leq \mu(R)\tilde{V} \\ m(R)|\xi|^2 \leq F_{p^i_\alpha p^j_\beta}(x,u,p)\xi^i_\alpha \xi^j_\beta \leq M(R)|\xi|^2 \qquad \forall\, \xi;\ m(R) > 0 \\ \tilde{V} = (1+|p|^2)^{1/2} \\ |u|^2 \leq R\,. \end{cases}$$

As it is clear, assumptions (II) do not grant the differentiability of the functional $J[u]$, but we have

THEOREM 5.2. *Assume that for* $|u| \leq R$

$$\nu\tilde{V}^2 - \lambda \leq F(x,u,p) \leq \mu(R)\tilde{V}^2$$

$$|F_u| \leq \mu(R)\tilde{V}^2$$

$$|F_p| \leq \mu(R)\tilde{V}\,.$$

Then the functional $J[u]$ *is differentiable in* $H^{1,2} \cap L^\infty(\Omega, R^N)$ *and its differential at* u *is still given by (5.3) where instead of* $H^{1,2}$ *one has to read* $H^{1,2} \cap L^\infty(\Omega, R^N)$.[19]

In the following we shall refer to assumptions (I) or (II) respectively as the *controllable growth conditions* and the *natural growth conditions*. And we say that u is an *extremal of the functional* $J[u]$ or a *weak solution to the Euler equation of the functional* $J[u]$ if

(a) *controllable growth conditions hold,* $u \in H^{1,2}(\Omega, R^N)$ *and satisfies*

$$(5.5) \qquad \int_\Omega [F_{p^i_\alpha}(x,u,\nabla u) D_\alpha \phi^i + F_{u^i}(x,u,\nabla u)\phi^i]dx$$

19) See [231] Theorem 1.10.3 for the proof.

for all $\phi \in H^1_0(\Omega, R^N)$, or

(b) *natural growth conditions hold,* $u \in H^{1,2} \cap L^\infty(\Omega, R^N)$ *and satisfies* (5.5) *for all* $\phi \in H^1_0 \cap L^\infty(\Omega, R^N)$.

More generally we shall deal in the following with *weak solutions* to nonlinear elliptic systems of the type

$$(5.6) \qquad \int_\Omega [A^\alpha_i(x, u, \nabla u) D_\alpha \phi^i + B_i(x, u, \nabla u)\phi^i] dx = 0 \qquad \forall \phi \in C^\infty_0(\Omega)$$

which in general are not Euler equations of a variational integral. The term 'weak solution' will have the same meaning as in (a) or (b) above with the formal change

$$A^\alpha_i \sim F_{p^i_\alpha}$$

$$B_i \sim F_{u^i}$$

and elliptic will mean that

$$A^\alpha_{i p^j_\beta} \xi^i_\alpha \xi^j_\beta \geq \nu |\xi|^2 \qquad \forall \xi;\ \nu > 0 .$$

We would like to remark that one can also consider systems with any polynomial growth condition, i.e.

$$|A^\alpha_i(x, u, p)| \leq c(|p|^m + \cdots)$$

and even higher order systems with any polynomial growth conditions; but in general we shall not do that in the following.

It is worth remarking that starting from the work of I. M. Višik [297] and Leray-Lions [199] there is a very large literature on the existence of weak solutions to nonlinear elliptic systems of the type (5.6) (at least in the case of controllable growth conditions:[20) see for example [204]

---

20) The same is not true in the case of natural growth conditions.

Finally we refer to [192] for weaker conditions under which one can derive the first variation of a regular functional; and we mention that there are a few more papers dealing with this problem, see e.g. [153] and its bibliography.

From now on we shall assume that a weak solution exists and we shall deal with the problem of regularity.

Chapter II

AN INTRODUCTION TO THE REGULARITY PROBLEM

As we have seen in Chapter I, by enlarging the space of competing functions we are able to prove the existence of generalized solutions to minimum problems for variational integrals, but we pay this simplicity with the new problem of the regularity of generalized solutions.

The aim of this chapter is to state what we mean by 'regularity problem', to give a short historical account and finally to present some counterexamples to the regularity for solutions of nonlinear elliptic systems or minimum points of variational integral defined on vector valued functions.

1. *Reduction to quasilinear and linear systems*

Let us first consider a stationary point of the simple regular integral

$$J[u] = \int_\Omega F(\nabla u)\,dx$$

or more generally a weak solution to the elliptic system

$$-D_\alpha A_i^\alpha(\nabla u) = 0 \qquad i = 1, \cdots, N$$

where $A_i^\alpha(p)$ satisfy the controllable growth condition $(I)_{bis}$, i.e. $A_i^\alpha \in C^1$

$$(1.1)\qquad \begin{cases} |A_i^\alpha(p)| \le c \cdot |p|\,, & |A^\alpha_{i\,p^j_\beta}(p)| \le L \\ A^\alpha_{i\,p^j_\beta}(p)\,\xi^i_\alpha \xi^j_\beta \ge \lambda|\xi|^2 & \forall \xi\,;\ \lambda > 0 \end{cases}$$

and $u \in H^{1,2}(\Omega, R^N)$ satisfies

$$\int_\Omega A_i^\alpha(\nabla u) D_\alpha \phi^i dx = 0 \qquad \forall \phi \in H_0^1(\Omega, R^N) . \tag{1.2}$$

First we want to show how the well-known quotient method will enable us to 'linearize' system (1.2).

Let us choose an integer s, $1 \leq s \leq n$, and set e_s for the unit vector in the x_s direction. If ϕ is a function with compact support in Ω and $|h|$ is small enough, we have

$$\int [A_i^\alpha(\nabla u(x+he_s)) - A_i^\alpha(\nabla u(x))] D_\alpha \phi^i dx = 0 .$$

Now for almost every x

$$A_i^\alpha(\nabla u(x+he_s)) - A_i^\alpha(\nabla u(x)) = \int_0^1 \frac{d}{dt} A_i^\alpha(t\nabla u(x+he_s) + (1-t)\nabla u(x))dt =$$

$$= \int_0^1 A^\alpha_{i_{p^j_\beta}}(t\nabla u(x+he_s)+(1-t)\nabla u(x))dt \cdot D_\beta [u^i(x+he_s) - u^i(x)]$$

hence, setting

$$\tilde{A}^{\alpha\beta}_{ij(h)}(x) = \int_0^1 A^\alpha_{i_{p^j_\beta}}(t\nabla u(x+he_s)+(1-t)\nabla u(x))dt$$

we get

$$\int \tilde{A}^{\alpha\beta}_{ij(h)}(x) D_\beta \frac{u^j(x+he_s) - u^j(x)}{h} D_\alpha \phi^i dx = 0 \tag{1.3}$$

and clearly

$$|\tilde{A}^{\alpha\beta}_{ij(h)}| \le L$$

$$\tilde{A}^{\alpha\beta}_{ij(h)} \xi^i_\alpha \xi^j_\beta \ge \lambda |\xi|^2 \qquad \forall \xi ; .$$

If we now insert in (1.3)

$$\phi = h^{-1}[u(x+he_s) - u(x)]\eta^2$$

where $\eta \in C^\infty_0(B_R)$, $B_{3R} \subset\subset \Omega$, $0 \le \eta \le 1$, $\eta \equiv 1$ on $B_{R/2}$, $|\nabla\eta| \le c/R$, we obtain

$$\int \left|\nabla \frac{u(x+he_s)-u(x)}{h}\right|^2 \eta^2 dx \le \text{const} \int \left|\frac{u(x+he_s)-u(x)}{h}\right|^2 |\nabla\eta|^2 \le$$

$$\le \frac{c}{R^2} \int_{B_{2R}} |\nabla u|^2$$

i.e.

$$\int_{B_{R/2}} \left|\nabla \frac{u(x+he_s)-u(x)}{h}\right|^2 dx \le \text{const independent of } h$$

which implies that $D_s u \in H^{1,2}(B_{R/4}, R^N)$ and therefore that $u \in H^{2,2}_{\text{loc}}(\Omega, R^N)$, by a standard-covering argument.[1] Moreover, passing

[1] Here we have used the well-known

PROPOSITION. *We have*

(a) *For each* $\Omega' \subset\subset \Omega$ *there exists a constant* $k = k(\Omega, \Omega')$ *such that if* $v \in H^{1,p}(\Omega)$ *and* $h < \frac{1}{2}\,\text{dist}(\Omega', \partial\Omega)$ *then*

$$\|\tau_{h,s} v\|_{L^p(\Omega')} \le k \|D_s v\|_{L^p(\Omega)} .$$

(b) *Let* $v \in L^p(\Omega)$, $1 < p < +\infty$. *Suppose that there exists a constant* k *such that for* $h < h_0$

$$\|\tau_{h,s} v\|_{L^p(\Omega')} \le k \qquad \Omega' \subset\subset \Omega$$

then $D_s v \in L^p(\Omega')$ *and*

$$\|D_s v\|_{L^p(\Omega')} \le k .$$

(c) *If* $v \in H^{1,p}(\Omega)$, $\Omega' \subset\subset \Omega$, *then* $\tau_{h,s} v \to D_s v$ *in* $L^p(\Omega')$. *Here*

$$\tau_{h,s} v = h^{-1}[v(x+he_s) - v(x)] .$$

to the limit for h going to zero in (1.3), we immediately deduce

$$\int_\Omega A^\alpha_{i p^j_\beta}(\nabla u) D_\beta(D_s u^j) D_\alpha \phi^i \, dx = 0 \qquad \forall \phi \in H^1_0(\Omega, R^N) \qquad \operatorname{spt} \phi \subset \Omega$$

which can be rewritten as a quasilinear system for the vector valued function

$$(1.4) \qquad U = (U^j_s) = (D_s u^j)_{\substack{s=1,\cdots,n \\ j=1,\cdots,N}}$$

as

$$(1.5) \qquad \int_\Omega \delta_{ls} A^\alpha_{i p^j_\beta}(U) D_\beta U^j_s D_\alpha \phi^i_l \, dx = 0 \qquad \forall \phi \in H^1_0(\Omega, R^{nN}) \qquad (\operatorname{spt} \phi \subset \Omega)\,.$$

Note that (1.5) is a *quasilinear elliptic* system, in fact

$$\delta_{sl} A^\alpha_{i p^j_\beta} \xi^{il}_\alpha \xi^{js}_\beta \geq \lambda \sum_{i,s=1}^{N} \sum_{\alpha=1}^{n} (\xi^{is}_\alpha)^2\,.$$

In conclusion we can state

THEOREM 1.1. *Let* $u \in H^1(\Omega, R^N)$ *be a weak solution to the elliptic system (1.2), where the* A^α_i *satisfy the controllable growth conditions. Then* $u \in H^{2,2}_{loc}(\Omega, R^N)$, *and the vector valued function* U *in (1.4) is a solution of the quasilinear elliptic system (1.5).*

The method we have used in deriving Theorem 1.1 is the well-known *difference quotient technique* see [247].[2)] Roughly speaking the idea is to differentiate the system

$$(1.6) \qquad -D_\alpha A^\alpha_i(\nabla u) = 0 \qquad i = 1, \cdots, N$$

[2)] See also [2] and [239] for the difference quotient technique applied to linear boundary value problems.

obtaining this way

$$-D_\alpha(A^\alpha_{i p^j_\beta}(\nabla u) D_\alpha D_s u^j) = 0 .$$

One has only to replace the ordinary derivatives with the difference quotient and use the proposition in footnote 1) Chapter II.

We would like to remark that in case of linear systems with smooth coefficients

$$(1.7) \qquad -D_\alpha(A^{\alpha\beta}_{ij}(x) D_\beta u^j) = 0 \qquad i = 1, \cdots, N, \qquad A^{\alpha\beta}_{ij} \in C^\infty$$

by differentiating, or more precisely by differencing, we get

$$-D_\alpha[A^{\alpha\beta}_{ij}(x) D_\alpha D_s u^j + A^{\alpha\beta}_{ijx_s} D_\beta u^j] = 0 \qquad i = 1, \cdots, N \qquad s = 1, \cdots, m$$

and then we can start again differentiating, and so on: this way one obtains that the weak solutions to system (1.7) belong to $C^\infty(\Omega)$. *This does not work in the nonlinear case (1.6)* (*we have to stop at the first step*) *and in the quasilinear case*

$$(1.8) \qquad \int_\Omega A^{\alpha\beta}_{ij}(u) D_\alpha u^i D_\beta \phi^j \, dx = 0 \qquad \forall \phi \in H^1_0(\Omega, R^N) .$$

This is formally clear, as a differentiation would give

$$\int_\Omega A^{\alpha\beta}_{ij}(u) D_\alpha(D_s u^i) D_\beta \phi^j dx + \int_\Omega A^{\alpha\beta}_{iju^\gamma}(u) D_s u^\gamma D_\alpha u^i D_\beta \phi^j dx = 0$$

and the second term on the left-hand side is not well defined for $\phi \in H^{1,2}_0(\Omega, R^N)$. But it is not even true that $u \in H^{2,2}_{loc}(\Omega, R^N)$. We shall see in Section 3 weak solutions to systems of the type (1.8) which do not belong to $H^{2,2}_{loc}(\Omega, R^N)$ (this is true, instead, for $N = 1$, see [191]).

Theorem 1.1 holds in general for nonlinear elliptic systems of the type

$$(1.9)\qquad \int_\Omega [A_i^\alpha(x,u,\nabla u)D_\alpha\phi^i + B_i(x,u,\nabla u)\phi^i]dx = 0 \qquad \forall\phi \in H_0^1(\Omega, R^N)$$

under *controllable growth conditions*. More precisely we have

THEOREM 1.1′. *Let* $u \in H^1(\Omega, R^N)$ *be a weak solution to the elliptic system (1.9), where* A_i^α *and* B_i *satisfy the controllable growth conditions. Then* $u \in H^{2,2}_{loc}(\Omega, R^N)$ *and the derivatives* $D_s u$ $s = 1, \cdots, n$ *verify*

$$(1.10)\qquad \int_\Omega [A^\alpha_{ip^j_\beta}(x,u,\nabla u)D_\beta D_s u^j + A^\alpha_{iu^j}(x,u,\nabla u)D_s u^j + A^\alpha_{ix_s}(x,u,\nabla u) +$$
$$+ \delta_{\alpha s}B_i(x,u,\nabla u)]D_\alpha\phi^i dx = 0$$

for all $\phi \in H_0^1(\Omega, R^N)$, spt $\phi \subset \Omega$.

The proof goes on as before, only a few technical complications will appear: we refer to [231].

We note that now it is not possible to read system (1.10) as a second order quasilinear elliptic system with respect to the unknown vector U as in (1.5); but, adding one more index as in (1.5), it can be seen as a fourth order quasilinear system in u.

Theorem 1.1′ does not hold in the case of *natural growth conditions*, even if $B \equiv 0$ and $A_i^\alpha = A_{ij}^{\alpha\beta}(u)D_\beta u^j$, as we have already stated. But we have (see [231][191][122] for further information):

THEOREM 1.2. *Let* $u \in H^{1,2} \cap L^\infty(\Omega, R^N)$ *be a weak solution to system (1.9),*[3] *where* A_i^α *and* B_i *satisfy the natural growth conditions. Then,*

---

[3] In (1.9) we now must read ... $\forall\phi \in H_0^1 \cap L^\infty(\Omega, R^N)$.

if moreover $u \in C^0(\Omega, R^N)$, *we have* $u \in H^{2,2}_{loc}(\Omega, R^N)$ *and (1.10) holds for all* $\phi \in H^1_0 \cap L^\infty(\Omega, R^N)$.

Proof. Let us assume that for R small the oscillation of u on B_R is sufficiently small. Putting as before $\phi = D_s(D_s u \cdot \eta^2)$ in (1.9), η^2 being a standard test function in B_R, $R < \text{dist}(x_0, \partial\Omega) \wedge R_0$, R_0 small, we easily get

$$\int |\nabla^2 u|\eta^2 dx \le c \int [(1 + |\nabla u|^2)\eta^2 + |\nabla u|^2 |\nabla \eta|^2 + |\nabla u|^4 \eta^2] dx$$

while putting $\phi = (u - u_R)|\nabla u|^2 \eta^2$ $u_R = \fint_{B_R} u = |B_R|^{-1} \int_{B_R} u dx$ we obtain

$$\int |\nabla u|^4 \eta^2 \le c \left\{ \int |\nabla u|^2 [\eta^2 + |\nabla \eta|^2 |u - u_R|^2] dx + \underset{B_R}{\text{osc}}\, u \cdot \int |\nabla^2 u|^2 \eta^2 \right\}.$$

Therefore, replacing η by η^2 and using the assumption, we deduce

$$\int |\nabla u|^4 \eta^4 + \int |\nabla^2 u|^2 \eta^4 \le \int [\eta^2 + |u - u_R|^4 |\nabla \eta|^4 + |\nabla u|^2 |\nabla \eta|^2 \eta^2] dx$$

which concludes the proof, apart from the fact that we should have used difference quotients instead of derivatives. q.e.d.

We shall refer to system (1.10) as *the system in variation* of system (1.9). Concluding we then have: *it is always possible to derive the system in variation under controllable growth conditions, while only continuous weak solutions are solutions to the system in variation if natural growth conditions hold.*

Let us now consider the system (1.9) and suppose for a moment that

(a) the functions A_i^α and B_i are smooth, say of class C^∞ or analytic

(b) the weak solutions to system (1.9) are functions of class $C^{1,\alpha}$.[4), 5)]

Thus we can read system (1.10) as a linear system with Hölder continuous coefficients and therefore rely on the linear theory for higher differentiability properties. To be more precise, let us consider system (1.6) and assume $A_i^\alpha \in C^\infty$ and $u \in C^{1,\alpha}$. Then for each fixed s, $1 \le s \le n$ the function $D_s u$ is a weak solution to

$$\int A_{ij}^{\alpha\beta}(x) D_\beta (D_s u^j) D_\alpha \phi^i dx = 0 \qquad \forall \phi \in H_0^1(\Omega, R^N)$$

with

$$A_{ij}^{\alpha\beta}(x) = A_{i\,p_\beta^j}^\alpha (\nabla u(x)) \in C^{0,\alpha}.$$

But from the linear theory (see Chapter III) we now know that $D_s u^i \in C_{loc}^{1,\alpha}(\Omega)$, hence $A_{ij}^{\alpha\beta}(x) \in C_{loc}^{1,\alpha}$ and then the linear theory again tells us that $D_s u^i \in C_{loc}^{2,\alpha}$, and so on.

We can now say that once we know that extremals, or weak solutions to nonlinear elliptic systems, are functions of class $C^{1,\alpha}$ (or of class $C^{0,\alpha}$ in the case of quasilinear systems like in (1.8)). Then higher regularity is a consequence of the linear theory for elliptic P.D.E.

But in general we are only able to find extremals or weak solutions in $H^{1,2}$ (or sometimes in $H^{1,2} \cap L^\infty$); therefore there is a gap in the regularity scale.

---

4) By $C^{0,\alpha}(\Omega)$ we denote the class of continuous functions satisfying a Hölder condition of order α, i.e.

$$[u]_{0,\alpha} = \sup_{x,y \in \Omega} |x-y|^{-\alpha} |u(x)-u(y)| < +\infty .$$

$C^{k,\alpha}(\Omega)$ denotes the class of $C^k(\Omega)$ functions whose k-derivatives are Hölder continuous with exponent α.

5) Then it is irrelevant to assume controllable or natural growth conditions.

The regularity problem (for nonlinear elliptic systems) is exactly the problem of filling this gap.

2. *A very brief historical note*

It would be very difficult to quote all the many different contributions to the regularity problem for linear and nonlinear elliptic systems. Here we shall confine ourselves to reporting about some fundamental steps mainly in connection with the nonlinear theory.

Probably one of the first results on the regularity problem for nonlinear equations is due to S. Bernstein [26] 1904, who proved that each solution of class C^3 of a nonlinear elliptic analytic second order equation in the plane $(N = 1, n = 2)$ is an analytic function. We must say that the analyticity of solutions of a single linear elliptic equation with analytic coefficients had already been proved by J. Hadamard in 1890.

Then different proofs were given of Bernstein's result; and in 1932 E. Hopf [169] proved an analogous analyticity theorem, still for one equation, but in arbitrary dimensions. In 1939 I. Petrowsky [253] proved the analyticity of solutions of a class of elliptic systems, and around 1957 C. B. Morrey, L. Nirenberg and A. Friedman gave the final contributions in order to prove the analyticity of solutions of general linear and nonlinear elliptic systems (see e.g. [231]). So we can state, as in the end of the first section: *every sufficiently smooth solution of a linear or nonlinear analytic elliptic system is an analytic function.*

Here we must recall at least the contributions by R. Caccioppoli [37] 1933,[6)] Schauder [260] 1934, A. Douglis, L. Nirenberg [73] 1954 and S. Agmon, A. Douglis, L. Nirenberg [3] 1959-64 to the regularity theory of linear systems.

Let us go back to the problem of the starting point. In 1912 L. Lichtenstein [202] proved that a C^2 extremal of a double regular integral is of class C^3 (and therefore analytic, if the equation is

6) See also [38] [39] [198] and particularly [40].

analytic); the same result was then proved by E. Hopf [168] in 1929 under the assumption that the solution be $C^{1,\alpha}$. And in fact, as we have seen at the end of Section 1, *in the above claim sufficiently smooth means* $C^{1,\alpha}$.[7)]

Actually in dimension $n = 2$ and for equations $N = 1$, C.B. Morrey was able to prove in 1938 Bernstein's result for Lipschitz solutions, solving this way Hilbert's problems we have stated in the introduction, in the case of the functionals considered by Haar and Radò (compare with Section 4 of Chapter I), see [221] and also [222]. But the assumption ' u is a Lipschitz solution' was not sufficient in general to prove regularity.

So no real progress was made on the regularity problem, except for the two dimensional case (1938-39 [220], see also [221][222]) where it happens that minimum points of quadratic functionals $(n = 2, N \geq 1)$ are Hölder continuous, until the celebrated result by E. De Giorgi [69] in 1957, see also J. Nash [237]. Let us illustrate it.

Let us consider the variational integral

$$J[u] = \int_\Omega F(\nabla u)\,dx \tag{2.1}$$

where $N = 1$, $n \geq 2$

$$m|p|^2 \leq F(p) \leq M|p|^2 \qquad m > 0$$

$$|F_p| \leq M|p|$$

$$\nu|\xi|^2 \leq F_{p_\alpha p_\beta}(p)\xi_\alpha\xi_\beta \leq M|\xi|^2 \qquad \forall\xi\,;\ \nu > 0$$

and let $u \in H^{1,2}(\Omega)$ be an extremal, i.e. a weak solution to

---

7) Really, C^1 is sufficient, see [225].

$$\int_\Omega F_{p_\alpha}(\nabla u) D_\alpha \phi dx = 0 \qquad \forall \phi \in H_0^1(\Omega) .$$

As we have seen in Section 1, the derivatives of u $D_s u$, $s = 1, \cdots, n$, are weak solutions of the equation

$$\int_\Omega F_{p_\alpha p_\beta}(\nabla u) D_\beta(D_s u) D_\alpha \phi dx = 0 \qquad \forall \phi \in H_0^1(\Omega) .$$

Now under the assumptions we have, we can only say that

$$F_{p_\alpha p_\beta}(\nabla u(x)) = A^{\alpha\beta}(x)$$

are measurable and bounded functions. So the regularity problem would be solved if we could show that weak solutions to *linear elliptic equations with* L^∞ *coefficients* are Hölder continuous: that this is true is exactly De Giorgi's result.

THEOREM 2.1 (De Giorgi). *Let* $u \in H^1(\Omega)$ *be a weak solution to*

$$\int_\Omega a^{\alpha\beta}(x) D_\alpha u D_\beta \phi dx = 0 \qquad \forall \phi \in H_0^1(\Omega)$$

where $a^{\alpha\beta}(x) \in L^\infty(\Omega)$ *and*

$$a^{\alpha\beta}\xi_\alpha \xi_\beta \geq \nu|\xi|^2 \qquad \forall \xi ; \nu > 0 .$$

Then $u \in C^{0,\alpha}_{\text{loc}}(\Omega)$ *for some positive* α, *and, for* $\tilde{\Omega} \subset\subset \Omega$, $\|u\|_{C^{0,\alpha}(\tilde{\Omega})} \leq c(\Omega, \tilde{\Omega}) \|u\|_{L^2(\Omega)}$.

By means of this theorem, the Hilbert problems of the introduction are completely solved for functionals of the type (2.1) with $N = 1$, n arbitrary:

in particular *if* F *in* (2.1) *is an analytic function, then each extremal of* (2.1) *is analytic.*

During the years 1959-60 C. B. Morrey, O. A. Ladyzhenskaya, N. N. Ural'tseva, G. Stampacchia stated analogous results for general linear equations with noncontinuous coefficients and even for nonlinear equations, proving that *weak solutions* (in the sense of Chapter I) *to elliptic nonlinear second order equations*, $N = 1$ (under controllable or natural growth conditions) *are smooth* (see [189][191], [231][283]). This way the regularity problem for one single equation can be considered as solved.

Besides a result by J. Nečas [238] for a class of higher order equations in dimension 2, no result was obtained during the years 1957-68 for the case $N > 1$. Many new proofs of De Giorgi's result were given (for example by Stampacchia, Moser, Landis, etc., see e.g. [129]) but none of these could be extended to cover the case of systems, although there was some hope in this sense.

In 1968 E. De Giorgi [71] showed that his result for equations cannot be extended to systems; in fact it does not hold for systems.

In the next section we shall present some examples which give a negative answer to the problem of regularity for systems.

3. *The vector valued case: some counterexamples to the regularity*

Let us start with De Giorgi's example.

EXAMPLE 3.1 (De Giorgi [71], 1968). Let Ω be the unit ball around the origin in R^N, $n \geq 3$. Consider the regular functional defined in $H^{1,2}(\Omega, R^N)$

$$(3.1)\quad J[v] = \int_\Omega \left\{ \sum_{\alpha, i=1}^{n} |D_\alpha v^i|^2 + \left[\sum_{\alpha, i=1}^{n} \left((n-2)\delta_{i\alpha} + n\frac{x_i x_\alpha}{|x|^2} \right) D_\alpha v^i \right]^2 \right\} dx$$

whose Euler equation is

$$(3.2) \qquad \int_\Omega A_{ij}^{\alpha\beta}(x) D_\beta u^j D_\alpha \phi^i \, dx = 0 \qquad \forall \phi \in H_0^1(\Omega, R^m)$$

with

$$A_{ij}^{\alpha\beta}(x) = \delta_{\alpha\beta}\delta_{ij} + \left[(n-2)\delta_{\alpha i} + n\frac{x_i x_\alpha}{|x|^2}\right]\left[(n-2)\delta_{\beta j} + n\frac{x_j x_\beta}{|x|^2}\right].$$

It is easily seen that $A_{ij}^{\alpha\beta} \in L^\infty(\Omega)$ and that there exist constants $0 < \nu \leq M$ such that

$$\nu|\xi|^2 \leq A_{ij}^{\alpha\beta}\xi_\alpha^i \xi_\beta^j \leq M|\xi|^2 \qquad \forall \xi .$$

Moreover one verifies that the vector valued function

$$u(x) = x \cdot |x|^{-\gamma} \qquad \gamma = \frac{n}{2}\{1 - [(2n-2)^2 + 1]^{-\frac{1}{2}}\}$$

which belongs to $H^{1,2}(\Omega, R^n)$ but is not bounded, is an extremal of the functional (3.1), i.e. a weak solution to (3.2); it is also the unique minimum point for $J[v]$ in (3.1) among the functions with vector x as prescribed value on the boundary.

In order to verify the above it is sufficient to note that system (3.2) is satisfied in a strong sense in $B_1(0)\setminus\{0\}$ and then use the following lemma,[8)] which is a slightly sharper version of a lemma due to De Giorgi.

LEMMA 3.1. *Let* Ω *be a bounded open set in* R^n, $n \geq 2$, $x_0 \in \Omega$, $u \in H^{1,2}(\Omega, R^N) \cap C^2(\Omega\setminus\{x_0\}, R^N)$ *and suppose that*

$$-D_\beta[A_{ij}^{\alpha\beta}(x, u, \nabla u) D_\alpha u^i] = f_j(x, u, \nabla u) \qquad 1 \leq j \leq N$$

in $\Omega\setminus\{x_0\}$, *where we assume that*

[8)] It will be the same for all counterexamples which follow.

$$A_{ij}^{\alpha\beta}(x, u(x), \nabla u(x)) \in C^1(\Omega\setminus\{x_0\}) \cap L^\infty(\Omega)$$

$$f_j(x, u(x), \nabla u(x)) \in C^0(\Omega\setminus\{x_0\})$$

and

$$|f(x, u, p)| \le a|p|^2 + b$$

for $x \in \Omega$, $u \in R^N$, $p \in R^{nN}$, a,b *constants*, $f = (f_1, \cdots, f_N)$. *Then it follows that*

$$\int_\Omega A_{ij}^{\alpha\beta}(x, u, \nabla u) D_\alpha u^i D_\beta \phi^j \, dx = \int_\Omega f_j(x, u, \nabla u) \phi^j dx$$

for all $\phi \in H^1 \cap L^\infty(\Omega, R^N)$ (*or in* $H_0^1(\Omega, R^N)$ *if* $f \equiv 0$).

We note that, since $x|x|^{-\gamma}$ is the gradient of the function $|x|^{2-\gamma}$ apart from a constant factor, it follows that $u(x) = |x|^{2-\gamma}$ is a weak solution to the fourth order elliptic equation

$$D_h D_k [A_{ij}^{hk}(x) D_i D_j u] = 0$$

$A_{ij}^{hk}(x)$ being the coefficients defined above. Remark that $|x|^{2-\gamma}$ is unbounded for $n \ge 5$.

We refer to [246] for a counterexample of the type 3.1 in connection with the theory of elasticity.

Example 3.1 shows that it is not possible to extend to systems De Giorgi's result in Theorem 2.1, and therefore to show regularity of the extremals of variational integrals in the same way as in the scalar case. But it leaves open the question whether weak solutions to quasilinear systems of the type

$$\int_\Omega A_{ij}^{\alpha\beta}(u) D_\alpha u^i D_\beta \phi^j \, dx = 0 \qquad \forall \phi \in H_0^1(\Omega, R^N)$$

with $A_{ij}^{\alpha\beta}$ smooth and

$$\nu|\xi|^2 \leq A_{ij}^{\alpha\beta}\xi_\alpha^i\xi_\beta^j \leq M|\xi|^2 \qquad \forall\xi\,;\ \nu > 0$$

can be regular. The following modification of De Giorgi's example due to E. Giusti and M. Miranda [138] gives a negative answer.

EXAMPLE 3.2 (Giusti-Miranda [138], 1968). Let Ω be the unit ball around the origin in R^n, $n \geq 3$. The vector valued function

$$u(x) = x|x|^{-1}$$

which belongs to $H^{1,2}(\Omega, R^n)$, is a weak solution to the elliptic system

$$\int_\Omega A_{ij}^{\alpha\beta}(u)\, D_\alpha u^i D_\beta \phi^j\, dx = 0 \qquad \forall\phi \in H_0^1(\Omega, R^m) \tag{3.3}$$

where

$$A_{ij}^{\alpha\beta}(u) = \delta_{ij}\,\delta_{\alpha\beta} + \left[\delta_{\beta j} + \frac{4}{n-2}\,\frac{u^j u^\beta}{1+|u|^2}\right]\left[\delta_{\alpha i} + \frac{4}{n-2}\,\frac{u^i u^\alpha}{1+|u|^2}\right]. \tag{3.4}$$

Note that the coefficients $A_{ij}^{\alpha\beta}(u)$ are real analytic functions in u. Moreover Giusti and Miranda prove that $x|x|^{-1}$ is, for n sufficiently large, the unique minimum point among functions $v \in H^1(\Omega, R^m)$ with $v = x$ on $\partial\Omega$ of the regular functional

$$\begin{aligned} J[v] = \int_\Omega F(v, \nabla v)\, dx = \int_\Omega \Bigg\{ &\sum_{i,j=1}^n |D_i v^j|^2 + \\ &+ \left[\sum_{i,j=1}^n \left(\delta_{ij} + \frac{4}{n-2}\,\frac{v^i v^j}{1+|v|^2}\right) D_i v^j\right]^2 \Bigg\}\, dx\,. \end{aligned} \tag{3.5}$$

Example 3.2 gives a final negative answer to the regularity problem for minima of functionals satisfying natural growth conditions. But, while it excludes the possibility of obtaining the regularity through general results for systems of the type of system in variation, leaves open the problem of the regularity of extremals of functionals of the type

$$J[u] = \int_\Omega F(\nabla u)\, dx .$$

Before considering this question, let us make a few remarks.

First note that system (3.3) is invariant under translations, hence also $(x-x_0)|x-x_0|^{-1}$ is a weak solution to system (3.3). Therefore we see that the singular set of solutions depends not only on the system but on the solution itself.

Since $x|x|^{-1}$ is the gradient of the function $v(x) = |x|$, then $v(x)$ is a weak solution of the fourth order quasilinear elliptic equation

$$D_h D_k(A_{ij}^{hk}(\nabla v) D_i D_j v) = 0$$

with coefficients A_{ij}^{hk} given by (3.4).

Simple modifications of Example 3.2 provide systems with unbounded solutions and solutions which do not belong to $H^{2,2}_{loc}(\Omega)$. The following example is taken from [244].

EXAMPLE 3.3. Set

$$A_{ij}^{\alpha\beta}(x,u) = \delta_{\alpha\beta}\delta_{ij} + c^2\left[\delta_{\alpha i} + b\,\frac{u^i u^\alpha |x|^{2\gamma-2}}{1+|u|^2|x|^{2\gamma-2}}\right]\left[\delta_{\beta j} + b\,\frac{u^j u^\beta |x|^{2\gamma-2}}{1+|u|^2|x|^{2\gamma-2}}\right]$$

where

$$\gamma \in \left[2, \frac{n}{2}\right) \qquad b = \frac{2n}{n-2} \qquad c^2 = \frac{\gamma(n-\gamma)(n-2)^2}{(n-2\gamma)^2(n-1)^2} .$$

Then $u(x) = |x|^{-\gamma}\cdot x$, which belongs to $H^{1,2}(B_1(0), R^n)$, is a weak solution of

$$\int_\Omega A_{ij}^{\alpha\beta}(x,u) D_\alpha u^i D_\beta \phi^j \, dx = 0 \qquad \forall \phi \in H_0^1(\Omega, R^n)$$

and obviously, for $\gamma \geq \frac{n-2}{2}$, it does not belong to $H^{2,2}_{loc}(\Omega)$.

We must mention that independently from Giusti-Miranda, analogous examples were provided by V. G. Maz'ja [208], and now different extensions are available, see for example J. Frehse [91], M. Giaquinta [105], S. A. Arakcheev [11].

Note that all counterexamples are in dimension $n \geq 3$; in fact, as we shall see, in dimension $n = 2$ we have regularity (at least in the case of controllable growth conditions).

Let us now come to the case of regular integrals of the type

$$J[u] = \int_\Omega F(\nabla u)\, dx \, . \tag{3.6}$$

EXAMPLE 3.4 (Nečas [241][242], 1975). J. Nečas presents a functional of the type (3.6), $n \geq 3$, $0 \in \Omega \subset R^n$, with F analytic, satisfying the growth conditions

$$|D^\ell F(p)| \leq c_1 \frac{|p|^2}{(1+|p|)^{|\ell|}} \qquad |\ell| = 0, 1, \cdots$$

where u is a vector with n^2 components, and the ellipticity condition

$$\frac{\partial^2 F}{\partial p_k^{ij}\, \partial p_\gamma^{\alpha\beta}} \xi_k^{ij} \xi_\gamma^{\alpha\beta} \geq c|\xi|^2 \qquad \forall \xi ;\ c > 0 \tag{3.7}$$

holds for n sufficiently large: it is a 2-times differentiable functional with definite positive second differential for n large. This functional has as extremal (and, hence, as minimum point in the class of functions with the same boundary value) the vector valued function $u^0(x)$ whose components are

$$u^{0ij} = \frac{x_i x_j}{|x|} .$$

Note that u^0 is a Lipschitz function, but it is not of class C^1. Nečas also presents a second order system of the type

$$\int_\Omega a_k^{ij}(\nabla u) D_k \phi_{ij}\, dx = 0 \qquad \forall \phi \in H_0^1(\Omega, R^{n^2}) \tag{3.8}$$

with analytic coefficients a_k^{ij}, satisfying for $n \geq 3$ the weaker ellipticity condition

$$a_k^{ij}(\xi)\xi_k^{ij} \geq c|\xi|^2 \qquad \forall \xi ; c > 0$$

and for $n \geq 5$ the ellipticity condition

$$\frac{\partial a_k^{ij}}{\partial p_\gamma^{\alpha\beta}} \xi_k^{ij} \xi_\gamma^{\alpha\beta} \geq c|\xi|^2 \qquad \forall \xi ; c > 0 \tag{3.9}$$

which has the same function $u^0(x)$ defined above as a weak solution. Nečas' functional is the following

$$\int_\Omega \left\{ \frac{1}{2} \frac{\partial u^{ij}}{\partial x_k} \frac{\partial u^{ij}}{\partial x_k} + \frac{\mu}{2} \frac{\partial u^{ij}}{\partial x_i} \frac{\partial u^{kk}}{\partial x_j} + \lambda \frac{\partial u^{ij}}{\partial x_i} \frac{\partial u^{ak}}{\partial x_a} \frac{\partial u^{1\beta}}{\partial x_1} \frac{\partial u^{jk}}{\partial x_\beta} (1 + |\nabla u|^2)^{-1} \right\} dx$$

with

$$\lambda = 2 \frac{n^3 - 1}{n(n-1)(n^3 - n + 1)}$$

$$\mu = -\frac{4 + n\lambda}{n^2 - n + 1} .$$

One sees by calculation that u^0 is an extremal and, noting that λ and μ go to zero when $n \to \infty$, one proves the ellipticity. We refer to the quoted papers for the other examples. We would like to mention also [126] [243] where one can find systems of the type (3.8) with C^∞ and analytic coefficients satisfying the strong ellipticity condition (3.9) for $n \geq 3$ and for which $u(x) = (u^{ij})$ where

$$u^{ij}(x) = \frac{x_i x_j}{|x|} - \frac{1}{n}\,\delta_{ij}|x|$$

is a weak solution.[9)]

It is worth remarking that we haven't any examples of a nonregular function $u : \Omega \subset \mathbf{R}^n \to \mathbf{R}^N$ $n = 3$, which is an extremal for a functional of type (3.6) satisfying the natural growth condition and the ellipticity condition.[10)]

However the counterexamples stated above are sufficient to say that weak solutions to nonlinear elliptic systems or extremals of regular integrals in the vector valued case are *non*smooth (in general).

The situation gets much worse when passing to consider quasilinear or nonlinear systems under natural growth conditions. As we have already seen, $H^{1,2}$ is not anymore the natural class for a weak solution to start with, and this even for a single equation $(N = 1)$ in two independent variables $(n = 2)$, see also [191]. Now we want to give some more justification of that.

Let us consider the equation

$$(3.10) \qquad -\Delta u = |\nabla u|^2$$

in the ball $B_R(0) = \{x : |x| < R\}$ $R < 1$, in $\mathbf{R}^2$, with the boundary condition

$$u(x) = 0 \quad \text{for} \quad |x| = R\,.$$

---

9) Note that all the counterexamples in 3.4 are invariant by translation.

10) It would be very interesting to have a counterexample of the type $u : \Omega \subset \mathbf{R}^3 \to \mathbf{R}^N$, $N = 2, 3$; compare also with [193].

This boundary value problem has the regular weak solution $u(x) \equiv 0$, as well as the irregular 'solution' $u(x) = \log\log|x|^{-1} - \log\log R \in H^{1,2}(B_R(0))$. (Compare with [191].) Hence we see that equation (3.10) violates the 'principle of local uniqueness' of weak 'solutions.'

The same happens in the case of Euler equations of regular functionals, as it has been noted by J. Frehse [93].

EXAMPLE 3.5 (Frehse [93], 1975). Let $n = 2$, $N = 1$, $r = e^{-1}$. The functional

$$J[u] = \int_{B_r(0)} [1 + (1 + e^u|\log|x||^{12})^{-1}]\,|\nabla u|^2\,dx$$

has $u(x) \equiv 0$ as minimum point in the class of functions with zero boundary value and $u(x) = 12\log\log|x|^{-1} \in H^{1,2}(B_r(0))$ as 'extremal.'

Therefore (in the case of natural growth conditions) we are led to consider definitively $H^{1,2} \cap L^\infty$ as the natural class where to start with weak solutions. But the most convincing argument in considering $H^{1,2} \cap L^\infty$ as the correct class is maybe the fact that, as we have already said, *weak solutions in* $H^{1,2} \cap L^\infty$ *of nonlinear equations,* $N = 1$, *under natural growth conditions are smooth,* see [190] and Chapters VII, IX of these notes.

For systems, i.e. $N > 1$, we cannot expect regularity under natural growths, compare with Example 3.2; but because of the quadratic growth on the right-hand side the situation becomes even worse, as shown by the two following examples due respectively to S. Hildebrandt, K.-O. Widman [164] and J. Frehse [92], see also E. Heinz [150].

EXAMPLE 3.6 (Hildebrandt-Widman [164], 1975). Let $n = N = 3$. The vector valued function

$$u(x) = x \cdot |x|^{-1}$$

is a weak solution to

$$-\Delta u = u|\nabla u|^2$$

and an extremal in $H^{1,2} \cap L^\infty$ for the functional

$$\int_\Omega a(|u|)\, |\nabla u|^2 \, dx$$

provided $a(t)$ is a smooth function with $a'(1) = -2a(1)$.

EXAMPLE 3.7 (Frehse [92], 1973). Let $n = N = 2$. The vector valued function $u(x) = (u^1(x), u^2(x))$ with

$$u^1(x) = \sin \log \log |x|^{-1} \qquad u^2(x) = \cos \log \log |x|^{-1}$$

which belongs to $H^{1,2} \cap L^\infty$, is a discontinuous weak solution of the system

$$\begin{cases} -\Delta u^1 = 2\, \dfrac{u^1 + u^2}{1 + |u|^2}\, |\nabla u|^2 \\[2ex] -\Delta u^2 = 2\, \dfrac{u^2 - u^1}{1 + |u|^2}\, |\nabla u|^2 \end{cases} .$$

Let us explicitly remark that in dimension $n = 2$ weak solutions of elliptic systems with natural growths may be irregular, and moreover that in Examples 3.6 and 3.7 the leading part is *diagonal*.

It is worth remarking that no variational counterexample (i.e. system which is the Euler equation of a regular functional) of the type in Example 3.7 is available, see [158] for a discussion (see also [142][258]).

Recently J. Frehse [96] has shown in dimension 2 a functional of the type

$$\int_\Omega F(x, u, \nabla u)\, dx$$

which has the vector valued function u in Example 3.7 as an extremal. Unfortunately the function $F(x, u, \eta)$ is analytic in (u, η) but only measurable in x.

CHAPTER III

LINEAR SYSTEMS: THE REGULARITY THEORY

In this chapter we shall present the Schauder-type estimates for linear systems in divergence form. These estimates are well known, see for example [3][73], but here we shall present a (maybe not so well-known) method, which appears in C.B. Morrey [225] and S. Campanato [45], to obtain them without using potential theory. This way we shall also state a few estimates we shall use in the following.

1. *An integral characterization of Hölder continuous functions*

Let $B(x_0, R)$ [1)] be the ball in $\mathbb{R}^n$ of radius R around x_0. The well-known Sobolev theorem states that if $u \in H^{1,p}(B_R(x_0))$ with $p > n$, then u is Hölder continuous with exponent $\alpha = 1 - \frac{n}{p}$. If $p \leq n$, u is not necessarily Hölder continuous.

For $x \in B_R(x_0)$, $0 < r \leq \delta(x) = R - |x - x_0|$, let us consider the non-increasing function

$$r \to \int_{B_r(x)} |\nabla u|^p \, dx \, . \tag{1.1}$$

The following classical result due to Morrey, see [231], states that if the function in (1.1) goes to zero fast enough uniformly in x, then u is Hölder continuous. More precisely we have

THEOREM 1.1 (Dirichlet growth theorem). *Let* $u \in H^{1,p}(B_R(x_0))$, $1 \leq p \leq n$. *Suppose that for all* $x \in B_R(x_0)$, *all* r, $0 < r \leq \delta(x) = R - |x - x_0|$

[1)] We shall also use the notation $B_R(x_0)$ or more simply, when no confusion may arise, B_R.

$$\int_{B_r(x)} |\nabla u|^p \, dx \le L^p \left(\frac{r}{\delta}\right)^{n-p+p\mu}$$

holds with $0 < \mu \le 1$. *Then* $u \in C^{0,\mu}(B_\rho(x_0))$ *for all* $\rho < R$; *moreover if* $|x-y| < \frac{\delta(x)}{2}$ *the following estimate holds*:

$$|u(x)-u(y)| \le c(n,p,\mu)\, L\, \delta^{1-\frac{n}{p}} \left[\frac{|x-y|}{\delta}\right]^\mu .$$

We refer to [231] Theorem 3.5.2 for the proof.

In this section we would like to prove a more general result, due to S. Campanato [43], see also N. G. Meyers [214], which implies Theorem 1.1 and characterizes Hölder continuous functions, see Theorem 1.2 below. This result will be very useful for studying the regularity of weak solutions to elliptic P.D.E. Although in the following we need only Theorem 1.2, we prefer to state it in the setting of the space $L^{p,\lambda}$ and $\mathcal{L}^{p,\lambda}$ defined below. In fact these spaces are very interesting in themselves and from time to time they will simplify our exposition.

Let Ω be a bounded connected open set in R^n and let us denote

$$\Omega(x,\rho) = \Omega \cap B(x,\rho)$$

$$\operatorname{diam} \Omega = \sup \{ |x-y| : x, y \in \Omega \} .$$

DEFINITION 1.1 (Morrey spaces). Let $p \ge 1$ *and* $\lambda \ge 0$. *By* $L^{p,\lambda}(\Omega)$ *we denote the linear space of functions* $u \in L^p(\Omega)$ *such that*

$$(1.2) \qquad \|u\|_{L^{p,\lambda}(\Omega)} = \left\{ \sup_{\substack{x \in \Omega \\ 0<\rho<\operatorname{diam}\Omega}} \rho^{-\lambda} \int_{\Omega(x,\rho)} |u|^p \, dx \right\}^{\frac{1}{p}} < +\infty .$$

It is easy to see that $\|u\|_{p,\lambda}$ in (1.2) is a norm respect to which $L^{p,\lambda}(\Omega)$ is a Banach space.

Obviously

$$\|u\|_p \le (\operatorname{diam}\Omega)^{\frac{\lambda}{p}} \|u\|_{L^{p,\lambda}(\Omega)}$$

$$\|u\|_{L^{p,0}(\Omega)} = \|u\|_{L^p(\Omega)}$$

hence $L^{p,0}(\Omega) \simeq L^p(\Omega)$.

Recalling that if $u \in L^1(\Omega)$ a.e.x_0 is a Lebesque point, i.e. for a.e.x_0

$$u(x_0) = \lim_{\rho \to 0^+} \frac{1}{|B(0,\rho)|} \int_{\Omega(x_0,\rho)} u(x)\,dx$$

by using Hölder inequality, we see that $L^{p,\lambda}(\Omega)$ reduces to the zero function for $\lambda > n$. Also from Hölder inequality we get

$$L^{q,\mu}(\Omega) \subset L^{p,\lambda}(\Omega) \quad \text{if} \quad \frac{n-\lambda}{p} \ge \frac{n-\mu}{q} \qquad p \le q\,.$$

Finally

$$\|u\|_{L^{p,n}(\Omega)} \le \omega_n^{\frac{1}{n}} \|u\|_\infty \qquad \omega_n = |B(0,1)|$$

$$|u(x_0)| \le \sup_\rho \frac{1}{\omega_n \rho^n} \left(\int_{\Omega(x_0,\rho)} |u|^p dx \right)^{\frac{1}{p}} |\Omega(x,\rho)|^{1-\frac{1}{p}} \qquad \text{a.e.} x_0$$

therefore

$$\|u\|_\infty \le \omega_n^{-\frac{1}{p}} \|u\|_{L^{p,n}(\Omega)}\,.$$

We can collect the properties just stated in the following proposition.

PROPOSITION 1.1. *We have*

a) $L^{p,0}(\Omega) \simeq L^p(\Omega)$

b) $L^{p,n}(\Omega) \simeq L^\infty(\Omega)$

c) $L^{p,\lambda}(\Omega) = \{0\}$ *for* $\lambda > n$

d) $L^{q,\mu}(\Omega) \subset L^{p,\lambda}(\Omega)$ *if* $p \le q$, $\frac{n-\lambda}{p} \le \frac{n-\mu}{q}$.

One could moreover show that

1. there exist functions $u \in L^{p,\lambda}(\Omega)$ $0 \leq \lambda < n$, which do not belong to any $L^q(\Omega)$ for $q > p$
2. $L^\infty(\Omega)$ is not a dense subspace in $L^{p,\lambda}(\Omega)$.

Set

$$u_{x_0,\rho} = \frac{1}{|\Omega(x_0,\rho)|} \int_{\Omega(x_0,\rho)} u(x)\,dx\,.^{2)}$$

DEFINITION 1.2 (Campanato spaces). *Let* $p \geq 1$ *and* $\lambda \geq 0$. *By* $\mathcal{L}^{p,\lambda}(\Omega)$ *we denote the linear space of functions* $u \in L^p(\Omega)$ *such that*

$$(1.3)\quad [u]_{p,\lambda} = \left\{ \sup_{\substack{x_0 \in \Omega \\ 0<\rho<\operatorname{diam}\Omega}} \rho^{-\lambda} \int_{\Omega(x_0,\rho)} |u(x) - u_{x_0,\rho}|^p\,dx \right\}^{\frac{1}{p}} < +\infty\,.$$

$\mathcal{L}^{p,\lambda}(\Omega)$ are Banach spaces with the norm

$$\|u\|_{\mathcal{L}^{p,\lambda}(\Omega)} = \|u\|_{L^p(\Omega)} + [u]_{p,\lambda}$$

and one sees that $u \in \mathcal{L}^{p,\lambda}(\Omega)$ if and only if

$$\sup_{\substack{x \in \Omega \\ 0<\rho<\operatorname{diam}\Omega}} \rho^{-\lambda} \inf_{c \in R} \int_{\Omega(x,\rho)} |u-c|^p\,dx < +\infty\,.$$

As in the case of Morrey spaces, using Hölder inequality, we get

$$\mathcal{L}^{q,\mu}(\Omega) \subset \mathcal{L}^{p,\lambda}(\Omega) \qquad p \leq q\,,\ \frac{n-\lambda}{p} \geq \frac{n-\mu}{q}\,.$$

---

2) When no confusion may arise, we shall also write u_ρ instead of $u_{x_0,\rho}$, with the meaning

$$u_{x_0,\rho} = u_\rho = \fint_{B_\rho} u\,dx = \frac{1}{|B_\rho|} \int_{B_\rho} u\,dx\,.$$

Since

$$\int_{\Omega(x_0,\rho)} |u-u_{x_0,\rho}|^p dx \le 2^{p-1}\left\{\int_{\Omega(x_0,\rho)} |u|^p dx + |\Omega(x_0,\rho)|\,|u_{x_0,\rho}|^p\right\}$$

and

$$|u_{x_0,\rho}|^p \le |\Omega(x_0,\rho)|^{-1}\cdot\int_\Omega |u|^p dx$$

we see that for $\lambda \le n$

$$[u]_{p,\lambda} \le 2\|u\|_{L^{p,\lambda}(\Omega)}$$

and therefore: *for* $0 \le \lambda \le n$ *we have* $L^{p,\lambda}(\Omega) \subset \mathcal{L}^{p,\lambda}(\Omega)$.

The following regularity condition on Ω leads us to state exactly the relation between Morrey and Campanato spaces.

DEFINITION 1.3. *Let* $A > 0$. *The bounded set* Ω *is said to be of type* (A) *if for all* $x_0 \in \Omega$ *and* $\rho < \operatorname{diam}\Omega$

$$|\Omega(x_0,\rho)| \ge A\rho^n .$$

This condition excludes that Ω may have sharp outward cusps; for instance all Lipschitz domains are of type (A) for some A.

We have

PROPOSITION 1.2. *Let* Ω *be of type* (A) *and* $0 \le \lambda < n$. *Then* $\mathcal{L}^{p,\lambda}(\Omega)$ *is isomorphic to* $L^{p,\lambda}(\Omega)$.

Proof. We have

$$(1.4)\quad \rho^{-\lambda}\int_{\Omega(x_0,\rho)} |u|^p dx \le 2^{p-1}\left\{\rho^{-\lambda}\int_{\Omega(x_0,\rho)} |u-u_{x_0,\rho}|^p dx + \omega_n\rho^{n-\lambda}|u_{x_0,\rho}|^p\right\}.$$

Hence, in order to prove the theorem, it is sufficient to estimate uniformly $\rho^{n-\lambda}|u_{x_0,\rho}|^p$ (in fact we already know that $L^{p,\lambda}(\Omega) \subset \mathcal{L}^{p,\lambda}(\Omega)$). For $0 < r < R$ we have

$$|u_{x_0,R} - u_{x_0,r}|^p \leq 2^{p-1}\{|u(x)-u_{x_0,R}|^p + |u(x)-u_{x_0,r}|^p\}$$

and integrating with respect to x on $\Omega(x_0, r)$

$$|u_{x_0,R} - u_{x_0,r}|^p \leq \frac{2^{p-1}}{Ar^n}\left[\int_{\Omega(x_0,R)} |u(x)-u_{x_0,R}|^p dx + \right.$$
$$\left. + \int_{\Omega(x_0,r)} |u(x)-u_{x_0,r}|^p dx\right]$$

from which the estimate

$$(1.5) \qquad |u_{x_0,R} - u_{x_0,r}| \leq c_1(p,A)\,[u]_{p,\lambda} R^{\frac{\lambda}{p}} r^{-\frac{n}{p}}$$

follows.

Set now $R_i = 2^{-i}\cdot R$; then (1.5) implies

$$(1.6) \qquad |u_{x_0,R_i} - u_{x_0,R_{i+1}}| \leq c_1 R^{\frac{\lambda-n}{p}}[u]_{p,\lambda} 2^{i\frac{n-\lambda}{p}+\frac{n}{p}}$$

which, taking the sum from 0 to h, gives

$$|u_{x_0,R} - u_{x_0,R_{h+1}}| \leq c_2(n,p,\lambda,A)\,[u]_{p,\lambda}\cdot R_{h+1}^{\frac{\lambda-n}{p}} .$$

Choosing now h and R with $\operatorname{diam}\Omega < R \leq 2\operatorname{diam}\Omega$, in such a way that $R_{h+1} = \rho$ we get

$$|u_{x_0,\rho}| \leq 2^{p-1}\{|u_{x_0,R}|^p + |u_{x_0,R} - u_{x_0,\rho}|^p\} \leq$$
$$\leq 2^{p-1}\{|u_{x_0,R}|^p + c_2^p \rho^{\lambda-n}[u]_{p,\lambda}^p\}$$

and finally from (1.4)

$$\rho^{-\lambda} \int_{\Omega(x_0,\rho)} |u|^p dx \leq \text{const}\ \{[u]^p_{p,\lambda} + |u_{x_0,R}|^p\} \leq \text{const}\ \|u\|^p_{\mathcal{L}^{p,\lambda}(\Omega)} .$$

q.e.d.

The spaces $\mathcal{L}^{p,\lambda}$ and $L^{p,\lambda}$ are not isomorphic for $\lambda \geq n$ (this is obvious for $\lambda > n$). In fact, for example, the function $u(x) = \log x$ belongs to $L^{p,1}(0,1)$ $\forall$ $p \geq 1$ but is not bounded.[3] Therefore

$$L^\infty(\Omega) \simeq L^{p,n}(\Omega) \subsetneq \mathcal{L}^{p,n}(\Omega) .$$

Let us now consider the case $n < \lambda \leq n + p$. We have

THEOREM 1.2 (An integral characterization of Hölder continuous functions). *Let* Ω *be of type* (A) *and* $n < \lambda \leq n + p$. *Then* $\mathcal{L}^{p,\lambda}(\Omega)$ *is isomorphic to the space* $C^{0,\alpha}(\Omega)$ *with* $\alpha = \frac{\lambda - n}{p}$. *Moreover if* $u \in \mathcal{L}^{p,\lambda}(\Omega)$ *with* $\lambda > n+p$, *then* u *is constant in* Ω.

Proof. Let $u \in C^{0,\alpha}(\Omega)$ and $x \in \Omega(x_0, \rho)$. We have

$$|u(x) - u_{x_0,\rho}| \leq 2^\alpha \rho^\alpha [u]_{C^{0,\alpha}(\Omega)}$$

and therefore

$$[u]_{p,\lambda} \leq \text{const}\ [u]_{C^{0,\alpha}(\Omega)} .$$

Now assume $u \in \mathcal{L}^{p,\lambda}(\Omega)$ with $\lambda > n$. For $R > 0$, set $R_i = 2^{-i}R$. For $k < h$, from (1.6) we get

$$|u_{x_0,R_k} - u_{x_0,R_h}| \leq \text{const} \cdot [u]_{p,\lambda} R_k^{\frac{\lambda-n}{p}} . \tag{1.7}$$

[3] Note that, if we argue as in the proof of Proposition 1.2, we see that the mean values of functions $u \in L^{p,n}(\Omega)$, although not equi-bounded, always blow up not faster than $|\log R|$, i.e.

$$|u_{x_0,R}| = 0(\log R) .$$

Therefore *the sequence* $\{u_{x_0,R_h}\}$ *is a Cauchy sequence for all* $x_0 \in \Omega$. Then set

$$\tilde{u}(x_0) = \lim_{h\to\infty} u_{x_0,R_h} .$$

Now we show that $\tilde{u}(x_0)$ *does not depend on the choice of* R. In fact for $r < R$ if we choose $j \geq i$ such that

$$R_{j+1} < r_i \leq R_j \qquad r_i = 2^{-i} \cdot r$$

using (1.7) and (1.5) we get

$$|u_{x_0,R_i} - u_{x_0,r_i}| \leq |u_{x_0,R_i} - u_{x_0,R_j}| + |u_{x_0,R_j} - u_{x_0,r_i}| \leq$$

$$\leq \text{const } [u]_{p,\lambda} [R_i^{\frac{\lambda-n}{p}} + R_j^{\frac{\lambda-n}{p}}] \leq \text{const } [u]_{p,\lambda} \cdot R_i^{\frac{\lambda-n}{p}} .$$

On the other hand $\{u_{x,\rho}\}$ converges, for $\rho \to 0^+$, in $L^1(\Omega)$ to the function u, so we have $u = \tilde{u}$ a.e., and going to the limit for $h \to \infty$ in (1.7), taking $k = 0$, we get

$$|u_{x,R} - u(x)| \leq \text{const } [u]_{p,\lambda} \cdot R^{\frac{\lambda-n}{p}} \tag{1.8}$$

that is, $\{u_{x,R}\}$ *converges uniformly to* $u(x)$ *in* Ω. Now since $x \to u_{x,R}$ are continuous functions, $u(x)$ *is continuous*. Finally we show that u is Hölder-continuous. Let $x, y \in \Omega$ and $R = |x-y|$. We have

$$|u(x) - u(y)| \leq |u_{x,2R} - u(x)| + |u_{x,2R} - u_{y,2R}| + |u_{y,2R} - u(y)| . \tag{1.9}$$

The first and third terms on the right-hand side of (1.9) are estimated in (1.8). For the second term we have

$$|u_{x,2R} - u_{y,2R}| \leq |u_{x,2R} - u(z)| + |u(z) - u_{y,2R}|$$

and integrating with respect to z over $\Omega(x, 2R) \cap \Omega(y, 2R)$

$$|u_{x,2R} - u_{y,2R}| \le |\Omega(x,2R) \cap \Omega(y,2R)|^{-1}$$

$$\left\{ \int_{\Omega(x,2R)} |u(z) - u_{x,2R}| dz + \int_{\Omega(y,2R)} |u(z) - u_{y,2R}| dy \right\}$$

i.e. using Hölder inequality

$$|u_{x,2R} - u_{y,2R}| \le \text{const } |\Omega(x,2R) \cap \Omega(y,2R)|^{-1} [u]_{p,\lambda} R^{\frac{\lambda-n}{p}+n} .$$

But $\Omega(x,2R) \cap \Omega(y,2R) \supset \Omega(x,R)$, hence $|\Omega(x,2R) \cap \Omega(y,2R)| \ge A \cdot R^n$. Therefore we finally get

$$|u(x) - u(y)| \le \text{const } [u]_{p,\lambda} \cdot R^{\frac{\lambda-n}{p}} = \text{const } [u]_{p,\lambda} |x-y|^{\frac{\lambda-n}{p}} .$$

In order to complete the proof, it remains to estimate $\sup u$. Let y be such that $u(y) = u_\Omega$; we have

$$|u(x)| \le |u_\Omega| + \text{const } [u]_{p,\lambda} (\text{diam } \Omega)^\alpha \le \text{const } \|u\|_{\mathcal{L}^{p,\lambda}(\Omega)} .$$

The second part of the theorem is now obvious. q.e.d.

Since in the following we deal mainly with local problems, we shall use Theorem 1.2 in the following weaker form:

THEOREM 1.3. *If*

$$\int_{B_\rho(x)} |u - u_{x,\rho}|^p dx \le c \rho^{n+p\alpha} \qquad \alpha \in (0,1]$$

for x *in an open set* Ω *and for all* $\rho < \min(R_0, \text{dist}(x, \partial\Omega))$ *(for some* R_0*), then* u *is locally Hölder-continuous with exponent* α *in* Ω.

REMARK 1.1. Because of Poincaré's inequality

$$\int_{B_\rho(x_0)} |u - u_{x_0,\rho}|^p dx \leq \text{const}\, \rho^p \int_{B_\rho(x_0)} |\nabla u|^p dx$$

it is clear that essentially Theorem 1.1 follows from Theorem 1.3.

REMARK 1.2. Theorem 1.2 permits to prove also Sobolev theorem: if $u \in H^{1,p}(\Omega)$ $p > n$, then $u \in C^{0,1-n/p}_{\text{loc}}(\Omega)$. In fact, by using Poincaré's inequality, we obtain

$$\int_{B_\rho(x_0)} |u - u_{x_0,\rho}| dx \leq \text{const} \cdot R \int_{B_\rho(x_0)} |\nabla u| dx \leq$$

$$\leq \text{const} \left(\int_{B_\rho(x_0)} |\nabla u|^p dx \right)^{\frac{1}{p}} R^{n - \frac{n}{p} + 1} .$$

We have not considered the case $\lambda = n$. Actually studying the space $\mathcal{L}^{p,n}(\Omega)$ requires deeper results. One could show in the case Ω = 'a cube of R^n' $= Q_0$ that $\mathcal{L}^{p,n}(\Omega)$, also called BMO (the space of functions with bounded mean oscillation), is isomorphic for all p to the so-called John-Nirenberg space $\mathcal{E}_0(\Omega)$, which can be defined in one of the following equivalent ways:

DEFINITION 1.4 (John-Nirenberg space). u *belongs to* $\mathcal{E}_0(Q_0)$ *if and only if*

a) *there exist two positive constants* H *and* β *such that*

$$\text{meas}\,\{x \in Q : |u(x) - u_Q| > \sigma\} \leq H e^{-\beta\sigma} |Q|$$

for all $\sigma > 0$ *and all cubes* Q *with edges parallel to the ones of* Q_0

b) *there exist two positive constants* k *and* M *such that for all cubes* Q *with edges parallel to the ones of* Q_0

$$\int_Q \exp(k|u-u_Q|)\,dx \le M|Q| .$$

We shall not prove this characterization, for which we refer to F. John, L. Nirenberg [178].

We end this section with a few bibliographic remarks and some extensions of the above results.

Special cases of Morrey spaces were introduced by Morrey already in 1938. The result due to John-Nirenberg [178], which probably is the deepest one in this field, is of 1961. But it is after 1960 that these spaces were introduced in a systematic way by S. Campanato [43] [44] [46] and studied by many authors, among others [49] [47], N. G. Meyers [214], G. Stampacchia [280] [281], J. Peetre [251], L. Piccinini [255] (see also the references of these papers). A systematic approach can be found in S. Campanato [53], A. Kufner, O. John, S. Fučik [187], E. Giusti [135] and J. Peetre [251] for what concerns mainly the interpolation theory.[4)]

Extensions and generalizations of these spaces are also available: we refer to the works quoted above and to their references. Here we confine ourselves to quote only one extension due to Campanato [44] and state an interpolation theorem [281] [47].

Let us denote by $\mathcal{P}_k$, k a nonnegative integer, the class of polynomial in x of degree $\le k$.

DEFINITION 1.5. $\mathcal{L}_k^{p,\lambda}(\Omega)$ $p \ge 1$, $\lambda \ge 0$, $k \ge 0$, *is the class of functions* $u \in L^p(\Omega)$ *such that*

$$[u]_{\mathcal{L}_k^{p,\lambda}(\Omega)} = \sup_{\substack{x\in\Omega \\ r>0}} \left[r^{-\lambda} \inf_{P\in\mathcal{P}_k} \int_{\Omega(x,r)} |u(y)-P(y)|^p\,dy \right] < +\infty .$$

---

4) Maybe it is worth remarking that $\mathcal{L}^{p,\lambda}(\Omega)$ are good spaces of interpolation: BMO is the dual space of Hardy's space, see Fefferman-Stein [88].

Then we have

a) *for* $\lambda > n + (k+1)p$, $\mathcal{L}_k^{p,\lambda}(\Omega) \equiv \mathcal{P}_k$

for $k = 1$

i) *if* $0 \le \lambda < n+p$, $\mathcal{L}_1^{p,\lambda}(\Omega) \simeq \mathcal{L}^{p,\lambda}(\Omega)$

ii) *if* $n+p < \lambda \le n+2p$, $\mathcal{L}_1^{p,\lambda}(\Omega) \simeq C^{1,\alpha}(\Omega)$, $\alpha = \dfrac{\lambda-(n+p)}{p}$

iii) *for* $\lambda = n+p$, $\mathcal{L}_1^{p,\lambda} \underset{\neq}{\supset} C^{0,1}(\Omega)$.

For $k \ge 1$ *we have*

i) *for* $0 \le \lambda < n+kp$, $\mathcal{L}_k^{p,\lambda}(\Omega) \simeq \mathcal{L}_{k-1}^{p,\lambda}(\Omega)$

ii) *for* $\lambda = n+kp$ *we have a limit space* $\mathcal{E}_k(\Omega)$

iii) *for* $n+kp < \lambda < n+(k+1)p$, $\mathcal{L}_k^{p,\lambda}(\Omega) \simeq C^{k,\alpha}(\Omega)$, $\alpha = \dfrac{\lambda-n}{p} - k$.

The following interpolation theorem due to G. Stampacchia [281] permits to avoid potential theory in studying the L^p-theory for linear elliptic systems, see [58] (see [123] for an application to the L^p-theory for stationary Stokes system).

THEOREM 1.4. *Let* Ω *be a bounded open set in* R^n *and* Q *a cube in* R^n. *Let*

$$T : L^p(\Omega) \to L^p(Q) \qquad \textit{for } p,\ 1 \le p < +\infty$$
$$T : \mathcal{E}_0(\Omega) \to \mathcal{E}_0(Q)$$

be linear and continuous with norms respectively M_1 *and* M_2. *Then* T *is linear and continuous from*

$$L^q(\Omega) \to L^q(Q)$$

for all q, $p < q < +\infty$ *and*

$$\|Tu\|_{L^q(\Omega)} \le k\|u\|_{L^q(\Omega)}$$

where $k = k(n, p, q, M_1, M_2, |\Omega|/_{|Q|})$.

2. *Linear systems with constant coefficients*

In this section we shall consider linear elliptic systems reduced to the leading part with constant coefficients[5)]

$$(2.1)\qquad -D_\alpha(A^{\alpha\beta}_{ij} D_\beta u^j) = 0 \qquad i = 1, \cdots, N\,.$$

Elliptic means that the coefficients satisfy the Legendre-Hadamard condition

$$(2.2)\qquad A^{\alpha\beta}_{ij}\, \xi_\alpha \xi_\beta \eta^i \eta^j \geq \nu|\xi|^2|\eta|^2 \qquad \forall\, \eta, \xi\,;\ \nu > 0$$

and we shall prove two simple estimates, see [45][225]. These two estimates (Theorem 2.1 below) will play a fundamental role in the whole regularity theory, and not only for linear systems.

Let us start by proving a very simple estimate to which we shall refer as *Caccioppoli inequality*:

PROPOSITION 2.1. *Let* $u \in H^1(\Omega, R^N)$ *be a weak solution to system (2.1), i.e.*

$$(2.3)\qquad \int_\Omega A^{\alpha\beta}_{ij} D_\beta u^j D_\alpha \phi^i dx = 0 \qquad \forall\, \phi \in H^1_0(\Omega, R^N)\,.$$

Then for all $x_0 \in \Omega$ *and all* $R < \frac{1}{2}\operatorname{dist}(x_0, \partial\Omega)$ *the following inequality holds*

$$(2.4)\qquad \int_{B_R(x_0)} |\nabla u|^2 dx \leq \frac{c}{R^2} \int_{B_{2R}(x_0)} |u|^2 dx\,.$$

As we shall see, Caccioppoli type inequalities hold for a large class of linear and nonlinear elliptic systems (see for example the beginning of Section 1 Chapter II and Chapter VI) and despite their simplicity, they are the starting point of the regularity theory: one could say that once

---

5) We shall confine ourselves to second order systems, but one can see that just by formal changes all results of this section (as well as of this chapter) extend to higher order systems.

Caccioppoli inequality holds then we have some result of regularity, although this is not completely true.

Proof. Although it is a very simple and standard one, let us give the proof (see Chapter II). Let η be a standard cut off function, i.e. $\eta \in C_0^\infty(B_{2R}(x_0))$, $0 \le \eta \le 1$, $\eta \equiv 1$ on $B_R(x_0)$, $|\nabla\eta| \le c/R$. Inserting $\phi = u\eta^2$ in (2.3) one immediately obtains

$$\int A_{ij}^{\alpha\beta} D_\beta(u^j\eta) D_\alpha(u^i\eta)\,dx \le c\left\{\int |u|\,|\nabla\eta|\,|\nabla(u\eta)| + \int |u|^2|\nabla\eta|^2\right\}.$$

Now by means of Fourier transform, we see that

$$\nu \int |\nabla(u\eta)|^2\,dx \le \int A_{ij}^{\alpha\beta} D_\beta(u^j\eta) D_\alpha(u^i\eta)\,dx$$

and the proof can be easily completed. q.e.d.

REMARK 2.1. Going into the proof of Proposition 1.1 (choosing η as before but with $\eta \equiv 1$ on B_ρ and $|\nabla\eta| \le c/_{R-\rho}$) and noting that if u is a solution to system 2.1 also $u-\lambda$, $\lambda = \text{const}$, is a solution, one sees immediately that Proposition 2.1 can be stated in the following stronger form: *if* u *is a solution to system (2.1), then for all* $x_0 \in \Omega$ *and for all* $\rho < R < \operatorname{dist}(x_0, \partial\Omega)$ *the following estimate holds*:

$$\int_{B_\rho(x_0)} |\nabla u|^2\,dx \le \frac{c}{(R-\rho)^2} \int_{B_R \setminus B_\rho} |u-\lambda|^2\,dx . \tag{2.5}$$

REMARK 2.2. By using the quotient method and Caccioppoli's estimate, as we have remarked in Chapter II, one can prove immediately higher regularity for weak solutions to system (2.1). For instance: *all weak solutions are* C^∞ *functions*; more precisely we have: *let* u *be a weak solution to system (2.1), then for all* $B_{R/2} \subset B_R \subset \Omega$ *and all* k

$$\|u\|_{H^k(B_{R/2})} \le c(k,R)\,\|u\|_{L^2(B_R)} \qquad (2.6)$$ [6)]

see e.g. [2][233].

We now state the two estimates:

THEOREM 2.1. *Let* u *be a weak solution to system (2.1). Then there exists a constant* c *depending on the constants of the system such that for each* $x_0 \in \Omega$ *and* $0 < \rho \le R \le \operatorname{dist}(x_0, \partial\Omega)$ *the following estimates hold*

$$\int_{B_\rho(x_0)} |u|^2 dx \le c\left(\frac{\rho}{R}\right)^n \int_{B_R(x_0)} |u|^2 dx \qquad (2.7)$$

$$\int_{B_\rho(x_0)} |u - u_{x_0,\rho}|^2 dx \le c\left(\frac{\rho}{R}\right)^{n+2} \int_{B_R(x_0)} |u - u_{x_0,R}|^2 dx\,. \qquad (2.8)$$

Proof. Let $\rho < R/2$ and $k > n$, using (2.6) and Sobolev imbedding theorem, we obtain

$$\int_{B_\rho(x_0)} |u|^2 dx \le c\rho^n \sup_{B_\rho(x_0)} |u|^2 \le c(R)\,\rho^n \|u\|_{H^k(B_{R/2})} \le$$

$$\le c(R)\,\rho^n \int_{B_R(x_0)} |u|^2 dx\,.$$

---

6) Let us remark that (2.6) holds also for elliptic systems with coefficients of class C^{k-1} with of course $c(k,R)$ depending also on the C^{k-1} norm of the coefficients and provided R is sufficiently small (depending on the modulus of continuity of coefficients). This can be seen by differentiating and freezing the coefficients in one point and then working as in the proof of the classical Gärding inequality. We refer again to [2][239].

Now it is easily seen, using a rescaling argument, that $c(R) = \text{const}\ R^{-n}$, i.e. (2.7) with $\rho < R/2$. Since (2.7) is obvious for $\rho \leq R/2$, we have (2.7) for $\rho < R$. Estimate (2.8) can be proved in the same way; or it is enough to note that the derivatives of u are also weak solutions, hence from (2.7)

$$\int_{B_\rho(x_0)} |\nabla u|^2 dx \leq c\left(\frac{\rho}{R}\right)^n \int_{B_R(x_0)} |\nabla u|^2 dx$$

and to use Poincaré inequality on the left-hand side and Caccioppoli inequality (2.5) on the right-hand side. q.e.d.

REMARK 2.3. It is worth remarking that estimates (2.7) and (2.8) hold for all derivatives of u, since all these derivatives are weak solutions of system (2.1).

Let us recall that *for* $u \in H^{m,p}(B_R(x_0))$ *there exists a unique polynomial* $P_{m-1}(x) = P_{m-1}(x_0, R, u; x)$ *of degree* $\leq m-1$ *such that*

$$\int_{B_R(x_0)} D^\alpha(u - P_{m-1})\, dx = 0$$

for all α, $|\alpha| \leq m-1$

$$P_{m-1}(x) = \sum_{|\alpha| \leq m-1} \frac{c_\beta}{\beta!}(x - x_0)^\beta$$

$$c_\beta = \sum_{2|\alpha| \leq m-1-|\beta|} c_{\beta,\alpha} R^{-n+2|\alpha|} \int_{B_R(x_0)} D^{\beta+2\alpha} u\, dx$$

with $c_{\beta,\alpha} = c_{\beta,\alpha}(n, m, \beta, \alpha)$.

Moreover the following *Poincaré type inequality* holds: *For every* $0 \leq s < t \leq m$ *there exists a constant* $c = c(n, p, s, t, m)$ *such that*

$$\int_{B_R(x_0)} \sum_{|\gamma|=s} |D^\gamma(u-P_{m-1})|^p\,dx \le$$

$$\le c\,R^{p(t-s)} \int_{B_R(x_0)} \sum_{|\gamma|=t} |D^\gamma(u-P_{m-1})|^p\,dx\,.$$

Then we immediately get

REMARK 2.4. Let u be a weak solution to system (2.1), then

$$\int_{B_\rho(x_0)} |u-P_{m-1}(x_0,\rho,u;x)|^2\,dx \le c\left(\frac{\rho}{R}\right)^{n+2m} \int_{B_R(x_0)} |u|^2\,. \tag{2.9}$$

Now we would like to prove a more precise estimate of the type (2.7) for harmonic functions. We have:

PROPOSITION 2.2. *Let* u *be a subharmonic function, then the function*

$$R \to R^{-n} \int_{B_R} u\,dx$$

is a nondecreasing function.

Since if u is harmonic u^2 (and $|\nabla u|^2$) is subharmonic, from Proposition 2.2 the estimate (2.7) follows with $c=1$.

Proof. From Gauss-Green formula it follows

$$0 \le \int_{\partial B_R(x_0)} \frac{du}{dn}\,d\sigma = R^{n-1}\frac{d}{dR}\int_{\partial B(0,1)} u(x_0+R\theta)\,d\theta$$

i.e. the function $R \to \int_{\partial B(0,1)} u(x_0+R\theta)\,d\theta$ is nondecreasing. Therefore

$$\int_{B_R(x_0)} u\,dx = \int_0^R \rho^{n-1}\,d\rho \int_{\partial B_1(0)} u(x_0+\rho\sigma)\,d\sigma \leq$$

$$\leq \int_0^R \rho^{n-1}\,d\rho \int_{\partial B(0,1)} u(x_0+R\sigma)\,d\sigma = \frac{R^n}{n} \int_{\partial B(0,1)} u(x_0+R\sigma)\,d\sigma =$$

$$= \frac{R}{n} \frac{d}{dR} \int_{B_R} u\,dx\,.$$

And now since $\phi(t) \leq \frac{t}{k}\,\phi'(t)$ implies that $t^{-k}\phi(t)$ is nondecreasing, we can conclude. q.e.d.

Let us prove a few results, which are well known for harmonic functions, and are a simple consequence of the established estimates.

Let us start by considering Caccioppoli inequality (2.5) and let us choose $\rho = R/2$, $\lambda = |B_R \setminus B_{R/2}|^{-1} \int_{B_R \setminus B_{R/2}} u\,dx$. Then we get

$$\int_{B_{R/2}} |\nabla u|^2\,dx \leq \frac{c}{R^2} \int_{B_R \setminus B_{R/2}} |u - u_{B_R \setminus B_{R/2}}|^2\,dx \tag{2.10}$$

and using Poincaré inequality

$$\int_{B_{R/2}} |\nabla u|^2\,dx \leq c \int_{B_R \setminus B_{R/2}} |\nabla u|^2\,dx\,.$$

Now filling the hole, see [300], i.e. adding c times the left-hand side we get

$$\int_{B_{R/2}} |\nabla u|^2 dx \le \theta \int_{B_R} |\nabla u|^2 dx \qquad \theta = \frac{c}{c+1} < 1 .$$

From which it follows immediately, for $R \to \infty$, that: *the only entire weak solutions to (2.1) with bounded Dirichlet integral are the constants.* This result follows immediately also from (2.8).[7)]

Now let us assume u bounded and $n = 2$; then from (2.10) it follows

$$\int_{B_{R/2}} |\nabla u|^2 dx \le \text{const independent of } R$$

therefore: *in dimension* 2 *the only bounded solutions to system (2.1) are the constants,*[8)] *compare also* [94][209][210].

It is worth remarking that these Liouville type theorems follow by using only Caccioppoli inequality, therefore they hold for instance for systems of the type

$$-D_\alpha[A_{ij}^{\alpha\beta}(x) D_\beta u^j] = 0 \qquad i = 1, \cdots, N$$

with coefficients $A_{ij}^{\alpha\beta} \in L^\infty$ and, satisfying the strong ellipticity condition

$$A_{ij}^{\alpha\beta} \xi_\alpha^i \xi_\beta^j \ge \nu|\xi|^2 \qquad \forall \xi ; \nu > 0 .$$

From estimate (2.9) we immediately derive: *if* u *is an entire weak solution of system (2.4) which grows polynomially at infinity, then it is a polynomial.*

---

7) More precisely we have: *if* $R^{-n} \int_{B_R} |\nabla u|^2 dx \to 0$ for $R \to +\infty$, *then* $u = \text{const}$.

8) Using the remark in the footnote 7) (this chapter): *if* u *is bounded in* R^n, $n \ge 2$, *then* $\int_{B_R} |\nabla u|^2 \le R^{-2} \int_{B_{2R}} |u|^2 \le cR^{n-2}$, *therefore* $u = \text{const}$.

Finally we would like to refer to [123] for extensions of the results of this section to systems of the type of stationary Stokes system.

In the following we shall need the maximum estimate stated below

PROPOSITION 2.3. *Let* u *be a weak solution in* B_R *to system (2.1) with* $u = \phi$ *on* ∂B_R *and* ϕ *bounded. Then* u *is bounded in* B_R *and*

$$\sup_{B_R} |u| \leq c \sup_{\partial B_R} |\phi|$$

with c *independent of* R.

This result is a consequence of a more general result, proved by means of a representation formula for the solution u, (see Canfora [59], see also [263]).

Actually it will be sufficient for us to have it in the following weaker form (cf. [60]).

PROPOSITION 2.4. *Let* $\phi \in H^{1,2} \cap L^\infty(B_R, R^N)$ *with* $\nabla\phi \in L^{2,n-2}(B_R, R^{nN})$, *and let* u *be a weak solution to system (2.1) with* $u - \phi \in H_0^1(B_R, R^N)$.
Then

$$\sup_{B_R} |u| \leq c\{\sup_{B_R} |\phi| + \|\nabla\phi\|_{L^{2,n-2}(B_R,R^{nN})}\}$$

with c *independent of* R.

Proof. Let $x \in B_R$ and

$$d = \operatorname{dist}(x, \partial B_R) = |x-y| \qquad y \in \partial B_R .$$

We have for all $0 < \rho < d$

$$\int_{B_\rho(x)} |u|^2 dx \leq c\left(\frac{\rho}{d}\right)^n \int_{B_d(x)} |u|^2 \leq c\left(\frac{\rho}{d}\right)^n \int_{B_{2d}(y)\cap B_R} |u|^2 \leq$$

$$\leq c\left(\frac{\rho}{d}\right)^n \left[d^n \sup_{B_R} |\phi|^2 + \int_{B(y,2d)\cap B_R} |u-\phi|^2 dx\right].$$

As Poincaré inequality is valid for $u - \phi$ in $B(y, 2d) \cap B_R$ (see for example [278], in fact the function $u - \phi$ is zero on a sufficiently large subset of $\partial B(y, 2d) \cap B_R$) we obtain

$$\frac{1}{\rho^n} \int_{B_\rho(x)} |u|^2 dx \leq c \left[\sup_{B_R} |\phi|^2 + d^{2-n} \int_{B_{2d}(y) \cap B_R} |\nabla(u-\phi)|^2 dx \right]$$

which concludes the proof, since from the global version of Theorem 2.2 below, see [45], also $\nabla u \in L^{2,n-2}(B_R)$. q.e.d.

Finally, let us consider linear nonhomogeneous elliptic systems with constant coefficients reduced to the leading part:

$$(2.11) \qquad -D_\alpha(A_{ij}^{\alpha\beta} D_\beta u^j) + D_\alpha f_i^\alpha = 0 \qquad i = 1, \cdots, N .$$

In order to illustrate the idea of C. B. Morrey [225] and S. Campanato [45] we want to prove now:

THEOREM 2.2. *Let us suppose* $f_i^\alpha \in \mathcal{L}^{2,\lambda}_{loc}(\Omega)$, $\lambda < n+2$, *and let* $u \in H^1_{loc}(\Omega, R^N)$ *be a weak solution to system (2.11). Then* $\nabla u \in \mathcal{L}^{2,\lambda}_{loc}(\Omega, R^{nN})$. *In particular if* $f_i^\alpha \in C^{0,\mu}_{loc}(\Omega)$ $i = 1, \cdots, N$ $\alpha = 1, \cdots, n$, *then the first derivatives of all weak solutions to (2.11) are Hölder continuous with exponent* μ.

Proof. Let $B_R(x_0) \subset\subset \Omega$ and let $v \in H^1(B_R(x_0), R^N)$ be the weak solution to the Dirichlet problem[9)]

$$\begin{cases} \int_\Omega A_{ij}^{\alpha\beta} D_\beta v^j D_\alpha \phi^i dx = 0 & \forall \phi \in H^1_0(B_R(x_0), R^N) \\ v - u \in H^1_0(B_R(x_0), R^N) . \end{cases}$$

9)The existence of such a function is an obvious consequence of the well-known Lax-Milgram theorem.

By Theorem 2.1 and Remark 2.3 we have for all $\rho < R$

$$(2.12) \qquad \int_{B_\rho(x_0)} |\nabla v - (\nabla v)_{x_0,\rho}|^2 dx \le c\left(\frac{\rho}{R}\right)^{n+2} \int_{B_R(x_0)} |\nabla v - (\nabla v)_{x_0,R}|^2 dx .$$

On the other hand, if we set $w = u - v$, we have $w = 0$ on $\partial B_R(x_0)$ and

$$\int_{B_R(x_0)} A_{ij}^{\alpha\beta} D_\beta w^j D_\alpha \phi^i dx = \int_{B_R(x_0)} (f_i^\alpha - f_{i,x_0,R}^\alpha) D_\alpha \phi^i \quad \forall \phi \in H_0^1(B_R, R^N) .$$

In particular we may take $\phi = w$, so that using the ellipticity relation and Hölder inequality we easily get

$$(2.13) \qquad \int_{B_R} |\nabla w|^2 dx \le c \int_{B_R} |f - f_{x_0,R}|^2 dx .$$

Now putting together (2.12) and (2.13) we obtain immediately

$$\int_{B_\rho(x_0)} |\nabla u - (\nabla u)_{x_0,\rho}|^2 dx \le c_1\left(\frac{\rho}{R}\right)^{n+2} \int_{B_R(x_0)} |\nabla u - (\nabla u)_{x_0,\rho}|^2 dx + \\ + c_2 \int_{B_R} |f - f_{x_0,R}|^2 dx$$

and because of the assumption on f

$$(2.14) \qquad \int_{B_\rho(x_0)} |\nabla u - (\nabla u)_{x_0,\rho}|^2 dx \le c_1\left(\frac{\rho}{R}\right)^{n+2} \int_{B_R(x_0)} |\nabla u - (\nabla u)_{x_0,R}|^2 dx + \\ + c_2 [f]_{\mathcal{L}^{2,\lambda}} \cdot R^\lambda .$$

Now the result would be a simple consequence of Theorems 1.2, 1.3 if in (2.14) we could write const $[f]_{\mathcal{L}^{2,\lambda}} \cdot \rho^\lambda$ instead of $c_2[f]_{\mathcal{L}^{2,\lambda}} \cdot R^\lambda$. This is in fact possible and is stated in the lemma below (cf. [45] [225] [123]).

q.e.d.

LEMMA 2.1. *Let* $\phi(t)$ *be a nonnegative and nondecreasing function. Suppose that*

$$\phi(\rho) \leq A\left[\left(\frac{\rho}{R}\right)^{\alpha} + \varepsilon\right]\phi(R) + BR^{\beta}$$

for all $\rho \leq R \leq R_0$, *with* A, α, β *nonnegative constants,* $\beta < \alpha$. *Then there exists a constant* $\varepsilon_0 = \varepsilon_0(A, \alpha, \beta)$ *such that if* $\varepsilon < \varepsilon_0$, *for all* $\rho \leq R \leq R_0$ *we have*

$$\phi(\rho) \leq c\left[\left(\frac{\rho}{R}\right)^{\beta}\phi(R) + B\rho^{\beta}\right]$$

where c *is a constant depending on* α, β, A.

Proof. For $0 < \tau < 1$ and $R < R_0$, we have

$$\phi(\tau R) \leq A\tau^{\alpha}[1 + \varepsilon\tau^{-\alpha}]\phi(R) + BR^{\beta} .$$

Choose now $\tau < 1$ in such a way that $2A\tau^\alpha = \tau^\gamma$ with $\alpha > \gamma > \beta$ and assume that $\varepsilon_0\tau^{-\alpha} < 1$. Then we get for every $R < R_0$

$$\phi(\tau R) \leq \tau^{\gamma}\phi(R) + BR^{\beta}$$

and therefore for all integers $k > 0$

$$\phi(\tau^{k+1}R) \leq \tau^{\gamma}\phi(\tau^k R) + B\tau^{k\beta}R^{\beta} \leq$$

$$\leq \tau^{(k+1)\gamma}\phi(R) + B\tau^{k\beta}R^{\beta}\sum_{j=0}^{k}\tau^{j(\gamma-\beta)} \leq$$

$$\leq c\,\tau^{(k+1)\beta}[\phi(R) + BR^{\beta}] .$$

Choosing k such that $\tau^{k+1}R < \rho \leq \tau^k R$, the last inequality gives at once (2.15).

q.e.d.

3. *Linear systems with continuous coefficients*

Let us consider linear elliptic systems with variable coefficients of the type

$$(3.1)\qquad -D_\alpha(A_{ij}^{\alpha\beta}(x)D_\beta u^j) + D_\alpha f_i^\alpha = 0 \qquad i = 1,\cdots,N\,.$$

We have:

THEOREM 3.1. *Suppose that* $A_{ij}^{\alpha\beta} \in C^0(\bar\Omega)$ $\alpha,\beta = 1,\cdots,n$, $ij = 1,\cdots,N$, $f_i^\alpha \in L^{2,\lambda}(\Omega)$ $0<\lambda<n$, *and let* u *be a weak solution to (3.1). Then* $\nabla u \in L^{2,\lambda}_{loc}(\Omega, R^{nN})$ *and for all* $\Omega_0 \subset\subset \Omega$

$$\|\nabla u\|_{L^{2,\lambda}(\Omega_0,R^{nN})} \le c(\Omega,\Omega_0)\{\|\nabla u\|_{L^2(\Omega,R^{nN})} + \|f\|_{L^{2,\lambda}(\Omega,R^{nN})}\}\,.$$ [10)]

Proof. We use the standard Korn's device of freezing the coefficients. Let $B_R(x_0) \subset\subset \Omega$, in $B_R(x_0)$ u is a weak solution to

$$-D_\alpha(A_{ij}^{\alpha\beta}(x_0)D_\beta u^j) + D_\alpha F_i^\alpha = 0$$

$$F_i^\alpha = f_i^\alpha + [A_{ij}^{\alpha\beta}(x_0) - A_{ij}^{\alpha\beta}(x)]D_\beta u^j\,.$$

Therefore as in the case of a system with constant coefficients using (2.7) for ∇u, we get

$$\int_{B_\rho(x_0)} |\nabla u|^2 dx \le c_1\Big(\frac{\rho}{R}\Big)^n \int_{B_R} |\nabla u|^2 + c_2 \int_{B_R} |f-f_R|^2 dx + c_3\omega^2(R) \int_{B_R} |\nabla u|^2 dx$$

where $\omega(R)$ is the modulus of continuity of the coefficients on $B_R(x_0)$:

$$\omega(R) = \sup_{B_R(x_0)} \Big\{\sum |A_{ij}^{\alpha\beta}(x) - A_{ij}^{\alpha\beta}(x_0)|^2\Big\}^{1/2}\,.$$

---

10) Of course $C(\Omega,\Omega_0)$ depends also on the modulus of continuity of the coefficients.

Now if R is smaller than some R_0, $c_3\omega^2(R)$ is small enough and we can use Lemma 2.1, obtaining the result immediately. q.e.d.

Note that if $f \in L^{2,n-2+2\mu}_{loc,\,0<\mu<1}$, then $\nabla u \in L^{2,n-2+2\mu}_{loc}$ and therefore the weak solution u is locally Hölder continuous with exponent μ.[11)]

THEOREM 3.2. *Suppose that* $A^{\alpha\beta}_{ij} \in C^{0,\mu}(\Omega)$ $ij = 1,\cdots,N$, $\alpha,\beta = 1,\cdots,n$, $f \in C^{0,\mu}(\Omega, R^{nN})$ $0 < \mu < 1$, *and let* u *be a weak solution to (3.1). Then* $\nabla u \in C^{0,\mu}_{loc}(\Omega, R^{nN})$ *and for all* $\Omega_0 \subset\subset \Omega$

$$[\nabla u]_{0,\mu,\Omega_0} \le c\{\|\nabla u\|_{L^2(\Omega,R^{nN})} + \|f\|_{C^{0,\mu}(\Omega,R^{nN})}\}.$$

Proof. As in Theorem 3.1, using (2.8) for ∇u we obtain for $B_\rho \subset B_R \subset\subset \Omega$

$$\int_{B_\rho} |\nabla u - (\nabla u)_\rho|^2 dx \le c_1\left(\frac{\rho}{R}\right)^{n+2} \int_{B_R} |\nabla u - (\nabla u)_R|^2 dx + \tag{3.2}$$

$$+ c_2 \int_{B_R} |f - f_R|^2 dx + c_3 \sup [A^{\alpha\beta}_{ij}]_{0,\mu} \cdot R^{2\mu} \int_{B_R} |\nabla u|^2$$

and, since we know from Theorem 3.1 that $\nabla u \in L^{2,n-\varepsilon}$ $\forall \varepsilon > 0$, we obtain

$$\int_{B_\rho} |\nabla u - (\nabla u)_\rho|^2 dx \le c_1\left(\frac{\rho}{R}\right)^{n+2} \int_{B_R} |\nabla u - (\nabla u)_R|^2 dx +$$

$$+ c_2 \int_{B_R} |f - f_R|^2 dx + c_4 R^{n+2\mu-\varepsilon}$$

i.e., using Lemma 2.1, that $\nabla u \in C^{0,\beta}_{loc}(\Omega)$, for all $\beta < \mu$. In particular

---

11) If $f^\alpha_i \in L^p_{loc}(\Omega)$ $p > n$, then $f^\alpha_i \in L^{2,n-2+2\mu}$ $\mu = 1 - \frac{n}{p}$.

∇u is locally bounded. We may then use inequality (3.2) again getting

$$\int_{B_\rho} |\nabla u - (\nabla u)_\rho|^2 \, dx \leq c_1 \left(\frac{\rho}{R}\right)^{n+2} \int_{B_R} |\nabla u - (\nabla u)_R|^2 \, dx + $$
$$+ c_5 R^{n+2\mu}$$

which concludes the proof, still through Lemma 2.1. q.e.d.

Theorems 3.1 and 3.2 are taken from S. Campanato [45]. They also appear in Morrey [225], the proof of Theorem 3.1 being essentially the same, while the proof of Theorem 3.2 is less transparent. Campanato's approach, which we have described, is instead more simple and useful. The method used here can be also used for showing regularity of weak solutions to complete (and even higher order) systems. With simple supplementary tricks it can be used for studying the boundary regularity (see [45] for the Dirichlet problem; but the same method works for example for the Neumann problem, see e.g. [123]). Then one is allowed to use the interpolation Theorem 1.4 (also locally) getting this way the L^p-theory for linear systems without potential theory, see [58][123].

Finally we would like to remark that the following result on higher order regularity can be easily deduced:

THEOREM 3.3. *Suppose that* $A_{ij}^{\alpha\beta} \in C^{k,\mu}(\Omega)$, $f \in C^{k,\mu}(\Omega, R^{nN})$, $0 < \mu < 1$, *and let* u *be a weak solution to (3.1). Then* $u \in C^{k+1,\mu}_{loc}(\Omega, R^N)$.

Chapter IV

SYSTEMS IN VARIATION: THE INDIRECT APPROACH TO THE REGULARITY

As we have seen in Chapter II, there is no hope of proving everywhere regularity for weak solutions to nonlinear elliptic systems, even in the simple case

(0.1) $$-D_\alpha[A^{\alpha\beta}_{ij}(x,u)D_\beta u^j] = 0 \qquad i = 1,\cdots,N .$$

The aim of this chapter is to present some *partial regularity* results for solutions of some nonlinear elliptic systems, essentially systems in variation of nonlinear systems satisfying controllable growth conditions.

These results are due to C. B. Morrey [232], E. Giusti, M. Miranda [139], E. Giusti [131] and are the starting point for the regularity theory for nonlinear systems. They read as: let u be a weak solution; then u is smooth in some open subset $\Omega_0 \subset \Omega$, and the singular set $\Omega \setminus \Omega_0$ is small.

Roughly speaking, the main idea of the proof is the following one: If some quantity, for instance $\displaystyle\fint_{B_R(x_0)} |u - u_{x_0,R}|^2\,dx$, that 'measures' the regularity in a neighborhood of x_0, is small enough, i.e. u varies very little for x near x_0, then the blowing up of u happens to converge to a solution of the 'tangent operator' which is a constant coefficient operator; therefore x_0 must be a regular point for u, as it is for the limit of the blown-up functions. This idea is very similar to the one used by E. De Giorgi [70] and J. F. Almgren [4] for proving the regularity of parametric minimal surfaces.

1. *Quasilinear systems: almost everywhere regularity*

In order to illustrate the main idea, let us start with the simple quasilinear system

$$(1.1)\qquad \sum_{i,j=1}^{N}\ \sum_{\alpha,\beta=1}^{n} \int_\Omega A_{ij}^{\alpha\beta}(x,u)\,D_\alpha u^i\, D_\beta \phi^j\,dx = 0 \qquad \forall\phi\in H_0^1(\Omega,R^N)$$

where $A_{ij}^{\alpha\beta}(x,u)$ are continuous functions satisfying

$$(1.2)\qquad |A_{ij}^{\alpha\beta}(x,u)| \le L$$

$$(1.3)\qquad A_{ij}^{\alpha\beta}\xi_\alpha^i\xi_\beta^j \ge |\xi|^2 \qquad \forall\xi\,.$$

For the sake of simplicity let us assume that the $A_{ij}^{\alpha\beta}$ *are uniformly continuous functions in* $\overline{\Omega}\times R^N$. Then we have, compare with [139]

THEOREM 1.1. *Let* $u\in H^1(\Omega,R^N)$ *be a weak solution to (1.1). Then there exists an open set* $\Omega_0\subset\Omega$ *such that* u *is Hölder continuous in* Ω_0, *and*

$$\operatorname{meas}(\Omega-\Omega_0) = 0\,.$$

The proof uses the following three facts and is essentially contained in the Main Lemma 1.1 below:

(a) Caccioppoli inequality: for $0<\rho<R<\operatorname{dist}(x_0,\partial\Omega)$

$$\int_{B_\rho(x_0)} |\nabla u|^2\,dx \le \frac{Q}{(R-\rho)^2}\int_{Q_R(x_0)} |u|^2\,dx$$

where $Q = Q(n,N,L)$.

(b) Estimate (2.8) Chapter III: let $b_{ij}^{\alpha\beta}$ be constants satisfying (1.2), (1.3); then there exists a constant $c = c(n,N,L)$ such that if $v\in H^1_{loc}\cap L^2(B_1(0),R^N)$ is a weak solution to

$$\int_{B_\rho} b_{ij}^{\alpha\beta} D_\alpha v^i D_\beta \phi^j \, dx = 0 \qquad \forall \phi \in H_0^1$$

then for all $0 < \rho < 1$ we have

$$U(0,\rho) \le c\,\rho^2 U(0,1)$$

where

$$U(x_0, R) = R^{-n} \int_{B_R(x_0)} |u - u_{x_0,R}|^2 \, dx \, .$$

(c) Let $a_{ij(h)}^{\alpha\beta}$ be a sequence of measurable functions satisfying (1.2), (1.3) and converging a.e. in $B(0,1)$ to $a_{ij}^{\alpha\beta}$ verifying (1.2) (1.3). Let u_h be a sequence in $H^1_{loc} \cap L^2(B_1(0), R^N)$ such that

(1.4) $$A_{(h)}(u_{(h)}, \phi) = \int a_{ij(h)}^{\alpha\beta}(x) D_\alpha u^i_{(h)} D_\beta \phi^j dx = 0 \qquad \forall \phi \in C^1_0(B(0,1), R^N)$$

and

$$u_{(h)} \to u \quad \text{weakly in} \quad L^2(B_1(0), R^N) \, .$$

Then $u \in H^1_{loc}(B_1(0), R^N)$ and for all $\rho < 1$

(1.5) $$u_{(h)} \to u \quad \text{strongly in} \quad L^2(B_\rho(0), R^N)$$

(1.6) $$\nabla u_{(h)} \to \nabla u \quad \text{weakly in} \quad L^2(B_\rho(0), R^{nN})$$

and moreover

(1.7) $$A(u, \phi) = \int a_{ij}^{\alpha\beta}(x) D_\alpha u^i D_\beta \phi^j \, dx = 0 \qquad \forall \phi \in C^1_0(B_1(0), R^N) \, .$$

Proof of (a) (b) (c): (a) It is sufficient to take, in (1.1), $\phi = u\eta^2$ with η the standard cutt-off function, compare with Section 2 in Chapter III.

(b) See Theorem 2.1 in Chapter III.

(c) Since the $u_{(h)}$ are equibounded in $L^2(B_1(0), R^N)$, from Caccioppoli inequality it follows that $\nabla u_{(h)}$ are equibounded in $L^2(B_\rho(0), R^N)$ for $\rho < 1$. Then (1.5) and (1.6) follow. Now we prove (1.7); we have

$$(1.8)\quad A_{(h)}(u_{(h)}, \phi) - A(u, \phi) = A(u_{(h)} - u, \phi) + A_{(h)}(u_{(h)}, \phi) - A(u_{(h)}, \phi)\,.$$

From (1.6) we obtain

$$A(u_{(h)} - u, \phi) \to 0 \quad \text{for} \quad h \to \infty$$

while if $\rho < 1$ and $\operatorname{spt} \phi \subset B_\rho(0)$

$$|A_{(h)}(u_{(h)}, \phi) - A(u_{(h)}, \phi)| \le \int |a_{(h)}(x) - a(x)|\,|\nabla u_{(h)}|\,|\nabla\phi|\,dx \le$$

$$\le c(\phi)\Big(\int_{B_\rho(0)} |a_{(h)} - a|^2\,dx\Big)^{1/2}\Big(\int_{B_\rho(0)} |\nabla u_{(h)}|^2\,dx\Big)^{1/2}$$

and therefore

$$A_{(h)}(u_{(h)}, \phi) - A(u_{(h)}, \phi) \to 0 \quad \text{for} \quad h \to \infty$$

which, through (1.8), finally gives

$$A_{(h)}(u_{(h)}, \phi) - A(u, \phi) \to 0\,.$$

This conclude the proof because of (1.4). q.e.d.

MAIN LEMMA 1.1. *For all* τ, $0 < \tau < 1$, *there exist two positive constants* $\varepsilon_0 = \varepsilon_0(\tau, n, N, L)$ $R_0 = R_0(\tau, n, N, L)$ *such that if* $u \in H^1(\Omega, R^N)$ *is a weak solution to (1.1) and for some* $x_0 \in \Omega$ *and some* $R < R_0 \wedge \operatorname{dist}(x_0, \partial\Omega)$ *we have*

$$(1.9)\qquad U(x_0, R) < \varepsilon_0^2$$

then

$$U(x_0, \tau R) \leq 2c\,\tau^2 U(x_0, R) \tag{1.10}$$

where c *is the constant in (b).*

Proof. Let us assume that the lemma is not true. Then there exist τ, $0 < \tau < 1$, a sequence $\{x_{(h)}\} \subset \Omega$, a sequence $\varepsilon_h \to 0$, a sequence $R_h \to 0$ and a sequence $u_{(h)} \in H^1(\Omega, R^N)$ of weak solutions to (1.1) such that

$$U^{(h)}(x_h, R_h) = \varepsilon_h^2$$

$$U^{(h)}(x_h, \tau R_h) > 2c\,\tau^2 \varepsilon_h^2 .$$

Translating x_h into the origin and blowing up, i.e. setting

$$v_{(h)}(y) = \varepsilon_h^{-1}[u_{(h)}(x_h + R_h y) - u_{(h)x_h, R_h}]$$

we have

$$\int_{B_1(0)} A_{ij}^{\alpha\beta}(x_h + R_h y, \varepsilon_h v^{(h)}(y) + u_{(h)x_h, R_h}) D_\alpha v^i_{(h)} D_\beta \phi^j dx = 0 \quad \forall \phi \in C_0^1(B_1(0), R^N)$$

$$V^{(h)}(0,1) = \int_{B_1} |v_{(h)}(y)|^2 dy = 1$$

$$V^{(h)}(0, \tau) > 2c\,\tau^2 . \tag{1.11}$$

Now, passing eventually to a subsequence, we have

$$v_{(h)} \to v \qquad \text{weakly in } L^2(B_1(0), R^N)$$

$$\varepsilon_h v_{(h)} \to 0 \qquad \text{a.e. in } B_1(0)$$

$$A_{ij}^{\alpha\beta}(x_h, u_{(h)x_h, R_h}) \to b_{ij}^{\alpha\beta}$$

and, hence, taking into account the uniform continuity of the $A_{ij}^{\alpha\beta}$, we have

$$A_{ij}^{\alpha\beta}(x_h + R_h y, \varepsilon_h v_{(h)} + u_{(h)x_h, R_h}) \to b_{ij}^{\alpha\beta} \quad \text{a.e. in} \quad B_1(0) .$$

From (c) then we get

$$\int_B b_{ij}^{\alpha\beta} D_\alpha v^i D_\beta \phi^j \, dx = 0 \qquad \forall \phi \in C_0^1(B_1(0), R^N)$$

and hence, because of the estimate in (b), we must have

$$V(0,\tau) \leq c\tau^2 V(0,1) .$$

On the other hand, using the semicontinuity of the norm in L^2 $V(0,1) \leq 1$, and because of (1.11) and (c)

$$V(0,\tau) > 2c\tau^2 .$$

Therefore we obtain a contradiction. q.e.d.

Proof of Theorem 1.1. Let $0 < \alpha < 1$ and choose τ in such a way that $2c\tau^{2-2\alpha} = 1$. Let $x_0 \in \Omega$ and $R < R_0 \wedge \operatorname{dist}(x_0, \partial\Omega)$ be such that

(1.12) $$U(x_0, R) < \varepsilon_0^2(\tau, n, N, L)$$

then we have from (1.10)

$$U(x_0, \tau R) < \tau^{2\alpha} U(x_0, R)$$

and hence

$$U(x_0, \tau R) < U(x_0, R) < \varepsilon_0^2 .$$

By induction, it follows for every integer k

$$U(x_0, \tau^k R) \leq \tau^{2\alpha k} U(x_0, R)$$

and hence for every $\rho < R$

$$U(x_0, \rho) \leq \text{const} \left(\frac{\rho}{R}\right)^{2\alpha} U(x_0, R) . \tag{1.13}$$

On the other hand, since $U(x_0, R)$ is a continuous function of x_0, if (1.12) holds for a point $x_0 \in \Omega$, then there exists a ball $B(x_0, r)$ such that for every $x \in B(x_0, r)$ we have

$$U(x, R) < \varepsilon_0^2 .$$

We then conclude that (1.13) holds uniformly for all x in $B(x_0, r)$, and therefore for every $x \in B(x_0, r)$

$$\int_{B_\rho(x)} |u(y) - u_{x,\rho}|^2 dx \leq \text{const } \rho^{n+2\alpha}$$

so that u is Hölder continuous in $B(x_0, r)$ with exponent α, $0 < \alpha < 1$, see Theorem 1.3, Chapter III. In conclusion, there exists an open set $\Omega_0 \subset \Omega$ such that the solution is locally Hölder continuous, with exponent α, in Ω_0. Note that the set Ω_0 is nonempty and independent of α, in fact $x_0 \in \Omega_0$ if and only if

$$\liminf_{R \to 0^+} \fint_{B_R} |u - u_{x_0,R}|^2 dx = 0 .$$

From that, since for almost every $x \in \Omega$

$$U(x, \rho) = \fint_{B_\rho(x)} |u - u_{x,\rho}|^2 dy \to 0$$

it follows also the estimate

$$\text{meas}(\Omega \setminus \Omega_0) = 0 .$$

q.e.d.

Let us remark explicitly that we have also proved that the singular set is given by

$$\Omega \setminus \Omega_0 = \left\{ x \in \Omega : \liminf_{R \to 0^+} \fint_{B_R(x_0)} |u - u_{x_0,R}|^2 dx > 0 \right\}.$$

We would like to note that Theorem 1.1 holds also under the weaker assumption that the coefficients $A_{ij}^{\alpha\beta}(x,u)$ be only continuous in $\Omega \times \mathbb{R}^N$, but in this case we have only

$$\Omega \setminus \Omega_0 \subset \left\{ x \in \Omega : \liminf_{R \to 0^+} \int_{B_R(x)} |u - u_{x,R}|^2 dx > 0 \right\} \cup$$

$$\cup \{ x : \sup_R |u_{x,R}| = +\infty \} .$$

More precisely we have

THEOREM 1.2. *For every* $M_0 > 0$, *there exist two positive constants* ε_0, R_0 *such that if* $u(x)$ *is a solution to system (1.1) in* Ω ($A_{ij}^{\alpha\beta}$ *being only continuous in* $\Omega \times \mathbb{R}^N$), *and if for some* $x_0 \in \Omega$ *and* $R < R_0 \wedge \operatorname{dist}(x_0, \partial\Omega)$

$$\int_{B_R(x_0)} |u|^2 dx < M_0^2 \;{}^{1)} \qquad \fint_{B_R(x_0)} |u - u_{x_0,R}|^2 dx < \varepsilon_0^2$$

then u *is Hölder continuous in a neighborhood of* x_0.

We omit the proof and we refer to [139] for it, since we shall see a different proof in Chapter VI.

The method described in this section applies to the study of more general systems; in Section 4 of this chapter we shall see some of the results that have been obtained.

---

1) Instead of $\int_{B_R(x_0)} |u|^2 dx < M_0^2$ one could require $|u_{x_0,R}| < M_0$.

We first want to improve the estimate of the measure of the singular set, precisely we want to present the result by E. Giusti [130] that permits to state:

$$\mathcal{H}^{n-2}(\Omega \setminus \Omega_0) = 0$$

where $\mathcal{H}^k$ is the k-dimensional Hausdorff measure. We shall do that in the next section, while in Section 3 we shall say something more about the singular set.

2. *The singular set*

Hausdorff measure. First let us rapidly recall the definition and a few properties of the Hausdorff measure. Let X be a metric space and J be a family of subset of X containing the empty set. Let

$$\zeta : J \to [0, +\infty]$$

be a function such that $\zeta(\phi) = 0$. Whenever $E \subset X$, we define

$$\mu_\varepsilon(E) = \inf \left\{ \sum_{h=0}^{\infty} \zeta(F_h) : F_h \in J, \bigcup_0^\infty F_h \supset E, \operatorname{diam} F_h < \varepsilon \right\} .$$

The fact that $\mu_\varepsilon \geq \mu_\delta$ for $0 < \varepsilon < \delta \leq +\infty$ implies the existence of

$$\mu(E) = \lim_{\varepsilon \to 0^+} \mu_\varepsilon(E) = \sup_{\varepsilon > 0} \mu_\varepsilon(E) \quad \text{whenever} \quad E \subset X .$$

The set function μ is called *the result of Carathéodory's construction from* ζ *and* J and it is usual to refer to μ_ε as *the size* ε *approximating measure.*

It is easy to verify that μ *is an outer measure for which the Borel sets are measurable.*

Appropriate choices of ζ and J yield measures μ of basic geometric importance; several such measures are defined and studied in H. Federer [85] to which we refer for more information. Here we are interested in the so-called k-*dimensional Hausdorff measure in* $\mathbb{R}^n$, which corresponds, for $X = R^n$, to choosing

$$\begin{aligned} &J = \text{the family of open sets in } R^n \\ (2.1) \qquad &\zeta(F) = \omega_k \cdot 2^{-k}(\operatorname{diam} F)^k \\ &\omega_k = \text{measure of the unit ball of } R^k \end{aligned}$$

and is denoted by $\mathcal{H}^k(E)$, i.e.

$$\mathcal{H}^k(E) = 2^{-k}\omega_k \sup_{\varepsilon>0} \inf \Big\{ \sum_{h=0}^{\infty} (\operatorname{diam} F_h)^k ;$$

$$\{F_h\} \text{ countable family of open sets, } \bigcup_{h=0}^{\infty} F_h \supset E, \operatorname{diam} F_h < \varepsilon \Big\} .$$

It is usual to define also $\mathcal{H}^0(E)$ as follows: whenever E contains a finite number of points, then $\mathcal{H}^0(E)$ is just the number of points of E, otherwise $\mathcal{H}^0(E) = +\infty$; $\mathcal{H}^0$ is called the *counting measure*.

One easily verifies that one obtains the same Hausdorff measure $\mathcal{H}^k$ by letting J be the family of closed subsets of R^n or the family of all subsets of R^n. Therefore $\mathcal{H}^k$ is a Borel regular measure.

With the same choice of ζ in (2.1), but letting J = the family of balls of R^n, the result of Carathéodory's construction is called the *k-dimensional spherical measure on* R^n, denoted by $\mathcal{S}^k$, and clearly

$$\mathcal{H}^k(E) \le \mathcal{S}^k(E) \le 2^k \mathcal{H}^k(E) .$$

But it happens that $\mathcal{H}^k \neq \mathcal{S}^k$ in general.[2)] Note anyway that the subsets of $\mathcal{H}^k$ measure zero coincide with the ones with zero $\mathcal{S}^k$ measure.

It is easily seen that if $\mathcal{H}^k(E)$ is finite then $\mathcal{H}^{k+\varepsilon}(E) = 0 \quad \forall \varepsilon > 0$; it is usual to define the *Hausdorff dimension of a set* E as

$$\dim_{\mathcal{H}} E = \inf \{k \in R^+ : \mathcal{H}^k(E) = 0\} .$$

[2)]See for example Besicowitch, Math. Ann. vol. 98 (1927), vol. 115 (1938), vol. 116 (1939).

Finally we have: i) the n-dimensional Hausdorff measure coincides with the outer n-dimensional Lebesgue measure, and therefore for a Lebesgue measurable set E

$$\mathcal{H}^n(E) = \mathcal{L}^n(E)$$

ii) subsets E with zero $\mathcal{H}^{n-1}$ measure do not disconnect Ω.

On the pointwise definition of $H^{1,p}$ *functions.* It is well known that L^1_{loc} functions can be defined almost everywhere as limit of their averages on balls. The following theorem gives an estimate of the dimension of Lebesgue points for $H^{1,p}$ functions:

THEOREM 2.1. *Let* Ω *be an open set of* R^n *and let* u *be a function belonging to* $H^{1,p}_{loc}(\Omega)$ $p < n$. *Set*

$$G = \{x \in \Omega : \nexists \lim_{\rho \to 0^+} u_{x,\rho}\} \cup \{x \in \Omega : \lim_{\rho \to 0^+} |u_{x,\rho}| = +\infty\} .$$

Then for all $\varepsilon > 0$

$$\mathcal{H}^{n-p+\varepsilon}(G) = 0$$

i.e. $\dim_{\mathcal{H}} G \leq n-p$.

In particular, if we choose in the equivalence class of u the function u^* defined for $x \notin G$ as

$$u^*(x) = \lim_{\rho \to 0^+} u_{x,\rho}$$

Theorem 2.1 permits to precise the pointwise value of $u \in H^{1,p}_{loc}$ except on a set whose Hausdorff dimension does not exceed $n-p$. Results of this kind can be found for example in M. Aronszajn et al. [12][13][14], $p = 2$, and in H. Federer [84], H. Federer, W. P. Ziemer [87] for functions u whose derivatives are measures, in particular $p = 1$. Here, for the proof of Theorem 2.1 as well as of Theorem 2.2 below, we follow E. Giusti [130].

The proof of Theorem 2.1 is based on the following result which has a relevant role in estimating the singular set of solutions of nonlinear elliptic systems.

THEOREM 2.2. *Let* Ω *be an open set of* R^n, v *be a function in* $L^1_{loc}(\Omega)$ *and* $0 \leq \alpha < n$. *Set*

$$E_\alpha = \left\{ x \in \Omega : \max \lim_{\rho \to 0^+} \rho^{-\alpha} \int_{B_\rho(x)} |v(y)| dy > 0 \right\}.$$

Then we have

$$\mathcal{H}^\alpha(E_\alpha) = 0 .$$

We need the following covering lemma.[3)]

LEMMA 2.1. *Let* A *be a bounded set in* R^n *and let* $r : x \to r(x)$ *be a function defined on* A *with range in* $(0,1)$. *Then there exists a sequence of points* $x_i \in A$ *such that*

$$\text{(2.2)} \qquad \begin{aligned} &B(x_i, r(x_i)) \cap B(x_j, r(x_j)) = \phi \quad \text{for} \quad i \neq j \\ &\bigcup_i B(x_i, 3r(x_i)) \supset A . \end{aligned}$$

Proof. Let us consider the family

$$B_{1,\frac{1}{2}} = \left\{ B(x, r(x)) : \tfrac{1}{2} \leq r(x) < 1 \right\} .$$

Since A is bounded, there exists a finite subfamily of disjoint balls

$$\overline{B}_{1,\frac{1}{2}} = \left\{ B(x_i, r(x_i)) : \tfrac{1}{2} \leq r(x_i) < 1 \quad i = 1, \cdots, n_1 \right\}$$

---

3) This lemma, together with Lemma 1.1 of Chapter V, gives a weak version of Besichovitch covering theorem.

which is maximal in the sense that each ball in $B_{1,1/2}$ intersects at least one element in $\overline{B}_{1,1/2}$. Once we have constructed $x_1, \cdots, x_{n_j}$, among the balls $B(x, r(x))$ with $2^{-j-1} \leq r(x) < 2^{-j}$ which do not intersect any of the balls $B(x_i, r(x_i))$, $i=1, \cdots, n_j$ we can find a finite family, say $n_{j+1} - n_j$ (eventually void), of balls such that each $B(x, r(x))$ with $2^{-j-2} \leq r(x) < 2^{-j-1}$ intersects at least one of the balls in $\{B(x_i, r(x_i)) : i=1, \cdots, n_{j+1}\}$. The sequence of the centers of these balls satisfies (2.2). In fact the balls $B(x_i, r(x_i))$ are disjoint by construction; for $x \in A$ there exists x_i such that

$$B(x, r(x)) \cap B(x_i, r(x_i)) \neq \phi$$

and $2r(x_i) \geq r(x)$. Hence

$$|x-x_i| \leq r(x) + r(x_i) \leq 3r(x_i)$$

and therefore $x \in B(x_i, 3r(x_i))$. q.e.d.

Proof of Theorem 2. It will be sufficient to show that for each compact subset $K \subset \Omega$

$$\mathcal{H}^{\alpha}(F) = 0 \quad \text{where} \quad F = E_\alpha \cap K .$$

Set

$$F^{(s)} = \left\{ x \in F : \max \lim_{\rho \to 0^+} \rho^{-\alpha} \int_{B(x,\rho)} |v(y)| dy > s^{-1} \right\}$$

obviously

$$F = \bigcup_{s=1}^{\infty} F^{(s)} .$$

Hence it will be sufficient to show that for all s

$$\mathcal{H}^{\alpha}(F^{(s)}) = 0 . \tag{2.3}$$

Let Q be a bounded open set with $K \subset Q \subset \overline{Q} \subset \Omega$ and $d = 1 \wedge \operatorname{dist}(x, \partial Q)$.

For all $\varepsilon > 0$, $0 < \varepsilon < d$, and for all $x \in F^{(s)}$ there exists $r(x)$, $0 < r(x) < \varepsilon$ such that

$$r(x)^{-\alpha} \int_{B(x,r(x))} |v(y)|dy \geq \frac{1}{2s} .$$

Let $\{x_i\}$ be the sequence in Lemma 2.1 (corresponding to $A = F^{(s)}$ $r_i = r(x_i)$). We have

$$\sum_i r_i^\alpha \leq 2s \sum_i \int_{B(x_i,r_i)} |v(y)|dy = 2s \int_{\cup B(x_i,r_i)} |v(y)|dy . \tag{2.4}$$

This inequality, since $\alpha < n$, implies

$$\begin{aligned} \text{meas}\,\{\cup B(x_i, r_i)\} &= \omega_n \sum_i r_i^n \leq \omega_n \varepsilon^{n-\alpha} \sum_i r_i^\alpha \leq \\ &\leq 2s\, \omega_n \varepsilon^{n-\alpha} \int_Q |v(y)|dy \qquad \omega_n = |B(0,1)| . \end{aligned} \tag{2.5}$$

From (2.5) and the absolute continuity theorem of Lebesgue (applied to (2.4)) it follows that the right-hand side of (2.4) goes to zero when $\varepsilon \to 0$; therefore taking into account (2.2) and the definition of $\mathcal{H}^\alpha(F^{(s)})$, (2.3) follows. q.e.d.

Proof of Theorem 1. Let us consider the subset E_α corresponding to the function $v = |\nabla u|^p$. In order to prove the theorem it will be sufficient to show that

$$G \subset E_{n-p+\varepsilon} \qquad \forall \varepsilon > 0 .$$

Fix $x_0 \in \Omega$, the function

$$r \to u_{x_0,r} = \omega_n^{-1} \int_{B(0,1)} u(x_0 + rx)\,dx$$

is a continuous function with continuous derivatives in the open interval $(0, \mathrm{dist}(x_0, \partial\Omega))$; and, since

$$\frac{d}{dr} u_{x_0,r} = \omega_n^{-1} \int_{B(0,1)} \sum_{i=1}^{n} x_i D_i u(x_0 + rx)\,dx$$

it follows

$$\left|\frac{d}{dr} u_{x_0,r}\right| \le \left(\omega_n^{-1} r^{-n} \int_{B(x_0,r)} |\nabla u|^p dx\right)^{1/p} . \tag{2.6}$$

For ε $0 < \varepsilon < p$, if $x_0 \notin E_{n-p+\varepsilon}$ we have

$$\sup_{0<r<r_0} \left\{\omega_n^{-1} r^{-n+p-\varepsilon} \int_{B(x_0,r)} |\nabla u|^p\right\}^{1/p} = L < +\infty$$

$$r_0 = 1 \wedge \frac{1}{2} \mathrm{dist}(x_0, \partial\Omega)$$

hence from (2.6)

$$\left|\frac{d}{dr} u_{x_0,r}\right| \le L r^{-1+\frac{\varepsilon}{p}}$$

and therefore, for $r, s < r_0$

$$|u_{x_0,r} - u_{x_0,s}| \le \left| \int_s^r \left| \frac{d}{dt} u_{x_0,t} \right| dt \right| \le \frac{Lp}{\varepsilon} \left| r^{\frac{\varepsilon}{p}} - s^{\frac{\varepsilon}{p}} \right|$$

which implies that $x_0 \notin G$. q.e.d.

In general it is not true that

$$\mathcal{H}^{n-p}(G) = 0$$

as the function $v(x) = (-\log|x|)^{1/4}$ shows.

The singular set $\Omega\setminus\Omega_0$. Let us now come to the singular set $\Omega\setminus\Omega_0$. As we have seen, weak solutions $u \in H^{1,2}(\Omega, R^N)$ to the elliptic system

$$-D_\alpha[A_{ij}^{\alpha\beta}(x,u)D_\beta u^j] = 0 \qquad i=1,\cdots,N$$

with continuous coefficients $A_{ij}^{\alpha\beta}$ are Hölder continuous in an open set Ω_0 (depending on the solution) and

$$\Omega\setminus\Omega_0 \subset \Sigma_1 \cup \Sigma_2$$

where

$$\Sigma_1 = \left\{ x \in \Omega : \liminf_{R\to 0^+} \fint_{B_R(x)} |u(y) - u_{x,R}|^2 dx > 0 \right\}$$

$$\Sigma_2 = \{x \in \Omega : \sup_R |u_{x,R}| = +\infty\} .$$

Note that because of Poincaré's and Caccioppoli's inequalities

$$\Sigma_1 = \left\{ x \in \Omega : \liminf_{R\to 0^+} R^{2-n} \int_{B_R(x)} |\nabla(y)|^2 dx > 0 \right\} .$$

Moreover, by adapting for example the argument in [213],[4] we can show that $u \in H^{1,p}_{loc}(\Omega)$ for some $p > 2$. Hence

$$R^{2-n} \int_{B_R(x)} |\nabla u|^2 \leq c \left(R^{p-n} \int_{B_R(x)} |\nabla u|^p \right)^{\frac{1}{p}} .$$

Therefore

$$\Sigma_1 \cup \Sigma_2 \subset E_{n-p} \cup G$$

---

[4] See anyway Chapter V.

where

$$E_{n-p} = \left\{ x : \max \lim_{\rho \to 0^+} \rho^{p-n} \int_{B_\rho(x)} |\nabla u|^p > 0 \right\}$$

$$G = \{x : \nexists \lim_{\rho \to 0^+} u_{x,\rho}\} \cup \{x : \lim_{\rho \to 0^+} |u_{x,\rho}| = +\infty\} .$$

And from Theorem 2.1 and 2.2 it follows that

$$\mathcal{H}^{n-p+\varepsilon}(\Omega \setminus \Omega_0) = 0 \qquad \forall \varepsilon > 0$$

where p is a suitable real number greater than 2. Note in particular that

$$\mathcal{H}^{n-2}(\Omega \setminus \Omega_0) = 0$$

i.e. the singular set is empty in dimension 2.[5)]

Therefore we can conclude:

THEOREM 2.3. *In Theorem 1.1 and in Theorem 1.2, for the singular set* $\Omega \setminus \Omega_0$, *we have*

$$\mathcal{H}^{n-q}(\Omega \setminus \Omega_0) = 0$$

for some $q > 2$.

All counterexamples in Chapter II Section 3 show solutions with singular set $\Omega \setminus \Omega_0$ = a point in dimension 3. From that, one can construct examples for which the singular set is a line in R^4, a plane in R^5 and so on. Therefore we could ask: is the dimension of the singular set of a weak solution to system (1.1) less or equal to $n-3$, i.e.

$$\mathcal{H}^{n-3+\varepsilon}(\Omega \setminus \Omega_0) = 0 \qquad \forall \varepsilon > 0 \, ?$$

This question, in its generality has no answer up to now; we refer to Chapter IX for a special case in which the answer is yes.

[5)] Note that, since $u \in H^{1,p}_{loc}(\Omega)$ for some $p > 2$, because of Sobolev theorem u is locally Hölder continuous in Ω.

3. *More on the regular and singular points*

Roughly speaking, Theorem 1.2 asserts that if u is sufficiently close to a constant vector in a sufficiently small ball, then it is regular near the center of the ball. On the other hand De Giorgi's theorem, Theorem 2.1 of Chapter II, implies that if a solution $u(x)$ of system (0.1) lies on a straight line

$$u(x) = \nu \cdot g(x) + \pi \qquad \pi \in R^N, \quad \nu \in S^{n-1} = \{x : |x| = 1\}$$

then u is regular, since it satisfies an elliptic equation.

Starting from this remark E. Giusti and G. Modica [140] have proved that if the vector $u(x)$ remains close to a straight line in a neighborhood of some point $x_0 \in \Omega$, then u is regular near x_0. More precisely

THEOREM 3.1. *For each* M_1 *there exist constants* ε_1 *and* R_1 *such that if* $u(x)$ *is a weak solution to system (1.1) and if for some* $x_0 \in \Omega$, $R < R_1 \wedge \operatorname{dist}(x_0, \partial\Omega)$, $\nu \in S^{n-1}$, $\pi \in R^N$, $|\pi| < M_1$ *we have*

$$\fint_{B_R(x_0)} |u|^2 dx \leq M_1^2$$

and

$$\fint_{B_R(x_0)} |u-\pi| dx - \fint_{B_R(x_0)} |(u-\pi, \nu)_{R^N}| < \varepsilon_1 .$$

then u *is regular in a neighborhood of* x_0.

Proof. We shall give only a sketch. The proof goes on as in Theorem 1.1, using De Giorgi's estimate (Theorem 2.1 Chapter II) instead of the estimate in (b) Section 1: if in (1.1) $A_{ij}^{\alpha\beta} = \delta_{ij} A^{\alpha\beta}(x)$, $A^{\alpha\beta} \in L^\infty$ then for the weak solutions $u = \nu \cdot g(x) + \pi$ we have

$$\|g\|_{C^{0,\delta}(B_{1/2})} \leq Q \|g\|_{L^2(B_1)} .$$

Then the idea is to reduce ourselves to applying Theorem 1.2, i.e. to prove that for every M_1 we can find ε_1 in such a way that

$$\fint_{B_{\tau R(x_0)}} |u|^2 \leq M_0^2 \qquad M_0 = 2M_1\sqrt{1+2Q}$$

and

$$\fint_{B_{\tau R}(x_0)} |u - u_{x_0,R}|^2 \leq \varepsilon_0^2$$

where M_0 and ε_0 are the constants in Theorem 1.2, provided τ is chosen suitably, and precisely as $\tau = \min\left\{\frac{1}{2}, \left(\frac{\varepsilon_0}{M_0}\right)^{1/\delta}\right\}$. If this were false, then $x_h, \pi_k, \nu_k, \varepsilon_k \downarrow 0$, $R_k \downarrow 0$ would exist such that

$$\fint_{B_{R_k}(x_k)} |u_k|^2 \, dx \leq M_1^2$$

(3.1)

$$\int_{B_{R_k}(x_k)} |u_k - \pi_k| dx - \int_{B_{R_k}(x_k)} |(u_k - \pi_k, \nu_k)| dx \leq \varepsilon_k$$

but either

$$\fint_{B_{\tau R_k}(x_k)} |u_k|^2 \geq M_0^2$$

or

$$\fint_{B_{\tau R_k}(x_k)} |u_k - u_{k x_k, R_k}|^2 > \varepsilon_0^2 .$$

Blowing up, i.e. considering

$$v_k(y) = u_k(x_k + R_k y)$$

one sees that $\{v_k\}$ converges to v, v satisfies the elliptic system

$$\int_B A_{ij}^{\alpha\beta}(x_0, v(y))\, D_\alpha v^i D_\beta \phi^j dx = 0 \qquad \forall \phi \in C_0^1(\Omega, R^N)$$

and

$$\fint_{B_\tau} |v|^2 \leq M_1^2$$

but either

(3.2) $$\fint_{B_\tau} |v|^2 dy \geq M_0^2$$

or

(3.3) $$\fint_{B_\tau} |v - v_{0,\tau}|^2 dy \geq \varepsilon_0^2 .$$

On the other hand passing to the limit in (3.1) one sees that for $\rho < 1$

$$\int_{B_\rho} (|v-\pi| - |(v-\pi, \nu)|)\, dy = 0 \qquad \rho < 1$$

so that

$$v(y) = \pi + (v(y), \nu)\, \nu = \pi + g(y)\, \nu$$

and we can conclude that

$$\fint_{B_\tau} |v|^2 \le \sup_{B_{1/2}} |g|^2 + M_1^2 = (2Q^2+1)M_1^2 = \frac{1}{4} M_0^2$$

$$\fint_{B_\tau} |v - v_{0,\tau}|^2 \le \fint_{B_\tau} |g - g_{0,\tau}|^2 \le 2Q^2 M_1^2 \tau^{2\delta} \le \frac{1}{4} \varepsilon_0^2 .$$

These two inequalities contradict (3.2)(3.3). q.e.d.

Now we would like to describe rapidly the connection between Liouville type theorems and regularity for Lipschitz solutions[6] of nonlinear systems

(3.4) $$- D_\alpha a_i^\alpha(x, u, \nabla u) + b_i(x, u, \nabla u) = 0 \qquad i = 1, \cdots, N .$$

This connection was pointed out by J. Frehse [94] and studied by M. Giaquinta and J. Necas [125][126], see also B. Kawohl [184].[7]

We say that *system (3.4) satisfies the Liouville property* (L) *if* $\forall x_0 \in \Omega$, $\forall \xi \in R^N$, *every solution* v *to system*

$$- D_\alpha a_i^\alpha(x_0, \xi, \nabla v) = 0$$

in R^n *with* $|\nabla v| \le c < +\infty$ *is a polynomial of at most first degree.*

Now we have

THEOREM 3.2. *Suppose* a_i^α, b_i *smooth. Denote by* K *the class of weak solutions* u *to (3.4) satisfying*

$$\|u\|_{H^{1,\infty}(\Omega, R^N)} \le c_1 .$$

---

[6] We recall that Lipschitz solutions are in general nonregular, see example 3.4 in Chapter II.

[7] Where bounded solutions to systems of the type in variation are considered. We refer also [209] [210] [212] and to [167] [159] for Liouville type theorems in a different context.

Assume that the Liouville property holds. Then for $u \in K$ *and for all* $\Omega' \subset\subset \Omega$

(3.5) $$\|u\|_{C^{1,\alpha}(\Omega', R^N)} \leq c_2(\Omega') .$$

In particular Liouville property implies regularity.

Proof. We give here an idea of the proof. The main point is to show that if $x_0 \in \Omega'$ then

(3.6) $$\nabla U(x_0, R) = \fint_{B_R(x_0)} |\nabla u - (\nabla u)_{x_0,R}|^2 dx \underset{R\to 0}{\to} 0$$

uniformly with respect to x_0 in Ω' and u in K; then the result follows as in Theorem 1.1, using the system in variation of system (3.4). In order to prove (3.6), set

$$u_R(y) = R^{-1}[u(x_0+Ry) - u(x_0)]$$

$x = x_0+Ry$ and set O_R for the image of Ω. Then we get

(3.7) $$\int_{O_R} [a_i^\alpha(x_0+Ry, Ru_R+u(x_0), \nabla_y u_R) D_\alpha \phi^i + b_i(x_0+Ry, Ru_R+u(x_0), \nabla_y u_R) R\phi^i] dx = 0 \quad \forall \phi \ \operatorname{spt} \phi \subset O_R$$

while the system in variation gives

(3.8) $$\int_{B_r(0)} |\nabla_y^2 u_R|^2 dx \leq c(r)$$

for $R < \frac{1}{2} \operatorname{dist}(\partial\Omega', \partial\Omega)$.

Now if (3.6) were false, there would exist $x_k \to x_0 \in \overline{\Omega}'$, $R_k \to 0$ and $u_{(k)} \in K$ such that

$$\nabla U^{(k)}(x_k, R_k) \geq \delta > 0 .$$

But, because of (3.8), we can suppose that $u_{(k)R_k} \to p$ in $H^{1,2}(B_R(0))$ $\forall r > 0$, and from (3.7) we get

$$\int_{R^n} a_i^{\alpha}(x_0, \xi, \nabla_y p)\, D_\alpha \phi^i\, dx = 0$$

where $\xi = \lim_{k\to\infty} u_k(x_0)$. Moreover $|\nabla p| \leq c < +\infty$, so that p is a polynomial of at most first degree. Hence

$$0 \to \int_{B_1(0)} |\nabla_y u_{(k)R_k} - \nabla p|^2 dy = R_k^{-n} \int_{B_{R_k}} |\nabla_x u_{(k)} - \nabla p|^2 dx \geq$$

$$\geq \nabla U^{(k)}(x_k, R_k)$$

gives a contradiction. q.e.d.

Moreover we have, see [173] see also [212], that regularity implies Liouville property; more precisely: *if* (3.5) *holds for all* $u \in K$, *then property* (L) *is true*. The proof we give here (see [212] for a different one) is based on the following argument, see [173]: Let v be a solution to

$$-D_\alpha a_i^\alpha(\nabla v) = 0 \qquad i = 1, \cdots, N$$

in R^n with $|\nabla v| \leq c < +\infty$, and assume $\operatorname{osc} \nabla v$ on $R^n = \omega > 0$. Now, also

$$v_R = \frac{v(x_0 + Rx)}{R}$$

are solutions with $|\nabla v_R| \leq c < +\infty$. However, since $\operatorname{osc} \nabla v_R$ in any neighborhood of the origin tends to ω as $R \to \infty$, there cannot be a common modulus of continuity for the functions ∇v_R in contradiction to (3.5). Therefore $\operatorname{osc} \nabla v = 0$, i.e. v is linear.

We refer to [126] for some applications. For instance from the fact that (L) implies regularity one can deduce that for $n \geq 1$ $N = 1$ or $n = 2$ $N \geq 1$ solutions are smooth, and that, under some explicit control on the ratio between the maximum and minimum 'eigenvalue' for the metric $\dfrac{\partial a^\alpha}{\partial p^j_\beta}$, solutions are also smooth (compare also with Koshelev [186]); for the optimality of the ratio in the last statement see [120].

4. *Systems in variation (controllable growth conditions): regularity results*

As we have already said, the first partial regularity result for solutions of nonlinear elliptic systems, $N > 1$, is due to C.B. Morrey [232]. He considers weak solutions $u = (u^1, \cdots, u^N)$, $u^i \in H^{m_i,k}(\Omega)$ $m_i \geq 1$, $i = 1, \cdots, N$ of elliptic nonlinear systems of higher order

$$(4.1) \qquad \sum_{i=1}^N \sum_{|\alpha| \leq m_i} \int_\Omega A^\alpha_i(x, Du) D^\alpha \phi^i \, dx = 0 \qquad \forall \phi^i \in C^\infty_0(\Omega)$$

where Du stands for $\{D^\alpha u^i\}$ $|\alpha| \leq m_i$; $i = 1, \cdots, N$, under controllable polynomial growth conditions, i.e.

$$(4.2) \quad \begin{cases} |A^\alpha_i(x,p)|, |A^\alpha_{ix_s}(x,p)| \leq MV^{k-1} \\ |A^\alpha_{ip^j_\beta}(x,p)|, |A^\alpha_{ip^j_\beta x_s}(x,p)| \leq MV^{k-2} \\ \displaystyle\sum_{ij=1}^N \sum_{\substack{|\alpha|=m_i \\ |\beta|=m_j}} A^\alpha_{ip^j_\beta} \pi^i_\alpha \pi^j_\beta \geq mV^{k-2}|\pi|^2 \qquad \forall \pi; \ m > 0 \\ \displaystyle V^2 = 1 + \sum_{i=1}^N \sum_{|\alpha| \leq m_i} (p^i_\alpha)^2 . \end{cases}$$

Let $k \geq 2$. As first step Morrey shows that under (4.2) it is possible to derive the system in variation, compare with Section 1, Chapter II and see also [231][297]. Precisely we have

THEOREM 4.1. *Let* u *be a weak solution to system (4.1), and (4.2) hold with* $k \geq 2$. *Then* $u^i \in H^{m_i+1,2}_{loc}(\Omega)$ $i = 1, \cdots, N$, $V^{k/2} \in H^{1,2}_{loc}(\Omega)$, *the vector* $p^i_\theta = u^i_{,x_\theta}$ [8] *satisfies*

$$\int_\Omega V^{k-2}[a^{\alpha\beta}_{ij}\, p^j_{\theta,\beta} + V e^{\theta\alpha}_i]\xi^i_{,\alpha}\, dx = 0 \qquad \forall \xi \in C^\infty_0(\Omega, R^N) \tag{4.3}$$

where the a's *and* e's *are bounded and measurable and*

$$V^{k-2} a^{\alpha\beta}_{ij}(x) = A^\alpha_{ip^j_\beta}(x, Du(x)) \qquad V^{k-1} e^{\theta\alpha}_i = A^\alpha_{ix_\theta}(x, Du(x)) .$$

Note that for $k = 2$, $m_i = 1$, $i = 1, \cdots, N$, system (4.3) reduces essentially to a system of type (1.1), compare with Chapter II.

Then Morrey proves the following

THEOREM 4.2. *Each* u^i *belongs to* $C^{m_i}_{loc}(\Omega_0)$ *where* Ω_0 *is an open subset of* Ω *and* $\mathcal{L}^n(\Omega \setminus \Omega_0) = 0$.

The idea of the proof is very similar to the one we presented when proving Theorem 1.1, although the proof is more involved.

Theorem 4.2, in case $k = 2$, $m_i = 1$, $i = 1, \cdots, N$, was proved for systems of the type (1.1), which include systems in variation (Theorem 1.1) by Giusti-Miranda [139] who also simplified the proof very much, and improved the estimate of the singular set as

$$\mathcal{H}^{n-1}(\Omega \setminus \Omega_0) = 0 . \tag{4.4}$$

Then, as we have seen, Giusti [130] improved (4.4), still for systems of the type (1.1), getting

---

8) Here we use the notation $u_{,x_\theta} = D_\theta u$ and more generally $u_{,\alpha} = D^\alpha u$.

$$\mathcal{H}^{n-p}(\Omega \setminus \Omega_0) = 0$$

for some $p > 2$.

The result of Theorem 4.2 was then extended, in the spirit of [139], to solutions of systems of the type (4.3).

More precisely, let us consider the quasilinear elliptic system

$$\int_\Omega V^{k-2} \sum_{i=1}^{N} \sum_{|\alpha| \leq m_i} \Big\{ \sum_{j=1}^{N} \sum_{|\beta| = m_j} A_{ij}^{\alpha\beta}(x, \delta u) D^\beta u^j + b_i^\alpha(x, \delta u) \Big\} D^\alpha \phi^i \, dx = 0 \qquad \forall \phi^i \in C_0^\infty(\Omega) \tag{4.5}$$

where δu stands for $D^\gamma u^h$ $h = 1, \cdots, N$, $|\gamma| \leq m_h$, and the coefficients $A_{ij}^{\alpha\beta}$ are continuous functions satisfying

$$|A_{ij}^{\alpha\beta}| \leq L$$

$$\sum_{ij=1}^{N} \sum_{\substack{|\alpha| = m_i \\ |\beta| = m_j}} A_{ij}^{\alpha\beta} \xi_\alpha^i \xi_\beta^j \geq |\xi|^2 \qquad \forall \xi$$

while

$$\sum_{i=1}^{N} \sum_{|\alpha| \leq m_i} |b_i^\alpha| \leq KV^2$$

where

$$V^2 = 1 + \sum_{i=1}^{N} \sum_{|\alpha| < m_i} |D^\alpha u^i|^2 .$$

Denote by $H^{m,k}(\Omega)$ the space of functions such that

$$V^{\frac{k-2}{2}} D^{\alpha} u^i \in L^2(\Omega) \qquad |\alpha| < m^i \quad i = 1, \cdots, N$$

$$V^{k-2} D^{\alpha} u^i \in L^2_{loc}(\Omega) \qquad |\alpha| = m_i \quad i = 1, \cdots, N .$$

Then we have, see [131][9] [132]

THEOREM 4.3. *Let* $k > 2$ *and* $u \in H^{m,k}$ *be a solution to system (4.5). Then there exists an open set* $\Omega_0 \subset \Omega$ *such that* $u^i \in C^{m_i-1,\alpha}(\Omega_0)$, $0 < \alpha < 1$, *and* $\mathcal{H}^{n-2+\varepsilon}(\Omega \setminus \Omega_0) = 0 \quad \forall \varepsilon > 0$.

Moreover

THEOREM 4.3′. *Under the hypotheses of Theorem 4.3, suppose that for some* p, $2 \le p \le n$, *we have either*

$$\int_{\Omega} V^{k-p} \sum_{i=1}^{N} \sum_{|\alpha|=m_i} |D^{\alpha} u^i|^p \, dx < +\infty \qquad 2 \le p \le k$$

or

$$\int_{\Omega} \sum_{i=1}^{N} \sum_{|\alpha|=m_i} |D^{\alpha} u^i|^p \, dx < +\infty \qquad k \le p \le m .$$

Then for all $\varepsilon > 0$ $\mathcal{H}^{n-p+\varepsilon}(\Omega \setminus \Omega_0) = 0$.[10]

Still assuming controllable polynomial $(k \neq 2)$ growth conditions, it remains for us to consider the case $1 < k < 2$.

In this case the quotient method seems not to work because $V^{k-2} < 1$, see [231], but Morrey [231] (Theorem 1.11.1″ and Section 5.10) show that one can derive the equation in variation provided u minimizes a regular

---

9)See also [63] for the boundary partial regularity for the Dirichlet problem.

10)See Chapter V where it describes one case in which this theorem applies.

functional and $m_i = m = 1 \ \forall i$. The same proof seems to work for the case $m_i = m > 1$, u being still a minimum point, but it is not clear what happens in the general case (4.1), $1 < k < 2$.

Anyway, we are allowed to consider the problem of the regularity for solutions of systems (4.5), $1 < k < 2$. The only result in this direction, still in the spirit of [139], is due to L. Pepe [252].

THEOREM 4.4. *Let* $u \in H^{1,k}(\Omega)$ *be a solution to system (4.5),* $m_i = 1$ $i = 1, \cdots, N$, $1 < k < 2$. *Then there exists an open set* $\Omega_0 \subset \Omega$ *such that* $u^i \in C^{0,\alpha}_{loc}(\Omega_0)$ $i = 1, \cdots, N$ *and* $\mathcal{H}^{n-k+\varepsilon}(\Omega \setminus \Omega_0) = 0$.

The extension of this result to the case $m_i > 1$ doesn't seem straightforward, and it is not even known whether the estimate of the singular set is 'optimal' or no.

All these results, even with the few gaps we have seen, give a good description of the regularity of weak solutions to nonlinear elliptic systems or extremals of regular functionals with *controllable* polynomial growth conditions. They also permit, as we have seen in Chapter II, to show higher regularity of course in the 'regular set' Ω_0 (which is connected if Ω is) provided the data are regular, compare with Chapter VI, Section 1.

Unfortunately the results given above do not apply if natural growth conditions hold,[11)] and the methods used to get them do not seem to carry over. In Chapter VI we shall present a different method, due to M. Giaquinta and E. Giusti [113] which permits to obtain some regularity results for nonlinear systems with natural but up to now only quadratic, polynomial growth. This method uses L^p-estimates for the gradient of the solutions which we shall prove in the following chapter.

Of course, the partial regularity results open new problems. We would like to close this section stating some of them:

[11)] With the exception of Theorems 4.1, 4.2 that hold also under natural conditions provided $m_i = 1$ $i = 1, \cdots, N$ and $k > n$, see [232]. For an analogous result for general systems of the type of systems in variation, see [57].

Singular set: There is the general problem of studying the singular set. In particular we can pose the following questions:

1. Is the singular set analytic or semianalytic?

2. Let us choose Σ in Ω with for example $\mathcal{H}^{n-2}(\Sigma)$ or $\mathcal{H}^{n-3+\varepsilon}(\Sigma) = 0 \quad \forall \varepsilon$, does an elliptic system exist with the solution having exactly Σ as singular set?

3. Are the singularities in dimension 3 isolated? (Compare with Chapter IX.)

Topological properties: among these

1. Is the regularity a generic property (with respect to the coefficients and/or the data)?

2. Is the class of systems with everywhere regular solutions dense, connected ... ?[12)]

And finally there is the general problem of *giving reasonable conditions for the solutions to be everywhere regular.*

[12)] It is not difficult to prove that the class of quasilinear system (1.1) whose solutions are smooth is open with respect to the uniform convergence of the coefficients, see the proof of Theorem 1.1, Chapter VI.

CHAPTER V

REVERSE HÖLDER INEQUALITIES AND L^p-ESTIMATES

Let us consider a weak solution u of the elliptic system

$$-D_\alpha[A_{ij}^{\alpha\beta}(x)D_\beta u^j] = 0 \qquad i = 1,\cdots,N \tag{0.1}$$

where $A_{ij}^{\alpha\beta} \in L^\infty(\Omega)$ and

$$A_{ij}^{\alpha\beta}\xi_\alpha^i\xi_\beta^j \geq |\xi|^2 \qquad \forall\xi;\ \nu > 0$$

then, as we know, we have the following Caccioppoli inequality: for $B_R \subset \Omega$

$$\int_{B_{R/2}} |\nabla u|^2 dx \leq \frac{c}{R^2}\int_{B_R} |u-u_R|^2 dx$$

and using Sobolev-Poincaré inequality

$$\int_{B_{R/2}} |\nabla u|^2 dx \leq \frac{c}{R^2}\left(\int_{B_R} |\nabla u|^q dx\right)^{2/q} \qquad q = \frac{2n}{n+2} < 2$$

which can be rewritten, dividing by R^n, as

$$\left(\fint_{B_{R/2}} |\nabla u|^2 dx\right)^{1/2} \leq c\left(\fint_{B_R} |\nabla u|^q dx\right)^{1/q} \text{ [1)]}. \tag{0.2}$$

---

1) $\fint_A f\,dx = \frac{1}{|A|}\int_A f\,dx$.

Except for the fact that integration is made on different sets, inequality (0.2) can be seen as a *reverse Hölder inequality.*

In Section 1 we shall prove a general theorem, implying in particular that from (0.2) higher integrability of the gradient of u follows, and in Section 2 we shall see some applications to nonlinear elliptic systems. In particular, in Section 2 we shall prove a few L^p-estimates, $p > 2$, for the gradient of weak solutions to nonlinear elliptic systems, which are interesting by themselves and will be one of the main tools in studying regularity in the next chapter.

Finally, in Section 3, an L^p-estimate for the gradient of minimum points of nondifferentiable functionals is proved.

1. *Reverse Hölder inequalities and higher integrability*

Roughly speaking, we shall state in this section that the reverse Hölder inequalities propagate in the exponents; more precisely we shall prove that a function g is L^p-integrable for some $p > q$ if the L^q-means of g over cubes do not exceed the L^1-means of g over suitable cubes for more than a fixed factor plus good terms, see Theorem 1.2 below.

Probably the first result in this direction is due to F. W. Gehring [103][2] (for more information see also [62] and the references there), and it is stated in terms of maximal functions.

Suppose $h \in L^1_{loc}(R^n)$, $h \geq 0$, *the maximal function*

$$M(h) : R^n \to [0, +\infty]$$

of h is defined by

$$M(h)(x) = \sup_R \int_{B_R(x)} h(y)dy$$

and we have:[3]

---

[2] In connection with the theory of quasiconformal mappings.

[3] We refer to E. M. Stein [285] for the proofs.

(a) *if* $h \in L^p(R^n)$, $1 \leq p \leq +\infty$, *then the function* $M(h)$ *is finite almost everywhere*

(b) *almost everywhere in* R^n $h(x) \leq M(h)(x)$

(c) *if* $h \in L^p(R^n)$ $1 < p \leq +\infty$, *then* $M(h) \in L^p(R^n)$ *and*

$$\|M(h)\|_{L^p(R^n)} \leq c(n,p)\|h\|_{L^p(R^n)} .$$

Let g be a nonnegative function defined on a cube Q of R^n, and think of g as zero out of Q,[4)] then Gehring's result [103] is the following:

THEOREM 1.1. *Suppose that almost everywhere on* Q

$$M(g^q) \leq b\,M(g)^q$$

where b *is a constant* > 1. *Then* $g \in L^p(Q)$ *for* $p \in [q, q+\varepsilon)$ *and*

$$\left(\fint_Q g^p\,dx\right)^{1/p} \leq c\left(\fint_Q g^q\,dx\right)^{1/q}$$

where ε *and* c *are positive constants depending only on* q, b *and* n.

But Theorem 1.1 is not yet useful for us. In fact, because of the restriction that g must be zero in $R^n \setminus Q$, (0.2) does not allow us to use Theorem 1.1. So we would like to have a *local* version of it. Now we come to state this local version.

Set

$$Q_R(x_0) = \{x \in R^n : |x_i - x_{0i}| < R \quad i = 1, \cdots, n\}$$

consider two nonnegative functions $g \in L^q(Q_1(0))$ $q > 1$, $f \in L^r(Q_1(0))$, $r > q$, and extend them equal to zero in $R^n \setminus Q_1(0)$. Denote by $d(x_0)$ the distance of $x_0 \in Q_1(0)$ from the boundary of $Q_1(0)$

---

4)This is essential for the validity of Theorem 1.1, compare with [103] and the appendix in [121].

For a nonnegative function $h(x) \in L^1_{loc}(R^n)$ define *the local maximal function* $M_{R_0}(h)$ *of* h *as*

$$M_{R_0}(h)(x) = \sup_{R<R_0} \fint_{B_R(x)} h(y)dy .$$

Then we have

THEOREM 1.2. *Suppose that almost everywhere on* $Q_1(0)$

$$M_{\frac{1}{2}d(x)}(g^q)(x) \le bM^q(g)(x) + M(f^q) + \theta M(g^q) \tag{1.1}$$

where $b > 1$ *and* $0 \le \theta < 1$. *Then* $g \in L^p_{loc}(Q_1(0))$ *for* $p \in [q, q+\varepsilon)$ *and*

$$\left(\fint_{Q_{1/2}(0)} g^p dx\right)^{\frac{1}{p}} \le c\left\{\left(\fint_{Q_1(0)} g^q dx\right)^{\frac{1}{q}} + \left(\fint_{Q_1(0)} f^p dx\right)^{\frac{1}{p}}\right\} \tag{1.2}$$

where $\varepsilon = \varepsilon(b, \theta, q, n)$ *and* $c = c(b, \theta, q, n)$ *are positive constants.*[5)]

As Theorem 1.2 is invariant by translations and dilatations, the following proposition follows immediately

PROPOSITION 1.1. *Let* Q *be an* n*-cube. Suppose*

$$\fint_{Q_R(x_0)} g^q dx \le c\left(\fint_{Q_{2R}(x_0)} g\, dx\right)^q + \fint_{Q_{2R}(x_0)} f^q dx + \theta \fint_{Q_{2R}(x_0)} g^q dx \tag{1.1$'$}$$

for each $x_0 \in Q$ *and each* $R < \frac{1}{2}\operatorname{dist}(x_0, \partial Q) \wedge R_0$, *where* R_0, b, θ *are constants with* $b > 1$, $R_0 > 0$, $0 \le \theta < 1$. *Then* $g \in L^p_{loc}(Q)$ *for* $p \in [q, q+\varepsilon)$ *and*

---

5) In fact they also depend on r. Moreover, on the right-hand side of (1.1) we may have $M_{\frac{1}{2}d(x)\wedge\tau_0}$, where τ_0 is a positive constant.

$$\left(\fint_{Q_R} g^p\,dx\right)^{\frac{1}{p}} \le c\left\{\left(\fint_{Q_{2R}} g^q\,dx\right)^{\frac{1}{q}} + \left(\fint_{Q_{2R}} f^p\,dx\right)^{\frac{1}{p}}\right\}$$

for $Q_{2R} \subset Q$, $R < R_0$, *where* c *and* ε *are positive constants depending only on* b, θ, q, n *(and* r *).*

Proposition 1.1 has been proved first, with θ small, by Giaquinta-Modica [121], Proposition 5.1. The idea of the proof is the same as the one by Gehring [103] plus a more refined covering argument of the type due to Calderon-Zygmund [42] stated in Lemma 1.3. Then Theorem 1.2 has been proved by E. W. Stredulinsky [287] in a slightly different way. The proof we present here is taken from M. Giaquinta [107] and is strictly related to the one in [121]; for a proof in a simpler situation see [111].

Let us remark that a higher integrability of $|\nabla u|$ follows immediately from (0.2) through Proposition 1.1 setting

$$g = |\nabla u|^{\frac{2n}{n+2}}$$

and $q = \frac{n+2}{n}$.

Before proving Theorem 1.2 we now state a few lemmas we shall use.

The first one is a covering lemma closely related to the more refined and well-known covering theorem by Vitali,[6)] see for example Stein [285]. (Compare also with Lemma 2.1 in Chapter IV, and footnote 3), Chapter IV.)

---

6)VITALI COVERING THEOREM. *Suppose that a measurable set* E *is covered by a collection of balls* $\{B_\alpha\}$, *in the sense that for each* $x \in E$, *and each* $\varepsilon > 0$, *there exists a* $B_{\alpha_0} \in \{B_\alpha\}$, *so that* $x \in B_{\alpha_0}$, *and* $\mathcal{L}^n(B_{\alpha_0}) < \varepsilon$. *Then there is a disjoint subsequence of these balls* $B_1, B_2, \cdots B_k, \cdots$ *so that*

$$\mathcal{L}^n(E - \bigcup_k B_k) = 0$$

see for example Saks [259] *or Federer* [85].

LEMMA 1.1. *Let* E *be a measurable subset of* R^n *which is covered by the union of a family of balls* $\{B_j\}$ *of bounded diameter. Then from this family we can select a disjoint subsequence* $B_1, B_2, \cdots$ *so that*

$$|E| \le c(n) \sum_k |B_k|$$

for example $c = 3^n$ *or* 5^n *will do.*

The second lemma is an obvious extension of an inequality for Stiltjes integrals proved, for example, in [103].

LEMMA 1.2. *Suppose that* $q \in (0, +\infty)$ *and* $a \in (1, +\infty)$, *that* $h(t), H(t):$ $[1, +\infty) \to [0, +\infty)$ *are nonincreasing functions with*

1. $\lim_{t\to\infty} h(t) = \lim_{t\to\infty} H(t) = 0$

2. $-\int_t^{+\infty} s^q\, dh(s) \le a[t^q h(t) + t^q H(t)] \qquad t \in [1, +\infty)$.

Then

$$-\int_1^{\infty} t^p\, dh(t) \le \frac{q}{aq-(a-1)p}\left(-\int_1^{\infty} t^q\, dh(t)\right) + \frac{a(p-q)}{aq-(a-1)p}\left(-\int_1^{\infty} t^p\, dH(t)\right)$$

for $p \in \left[q, \frac{a}{a-1}\, q\right)$.

Proof. Let us first assume that there exists $j \in (1, +\infty)$ such that

$$h(t) = 0 \quad \text{for} \quad t \in [j, +\infty)$$

and set

$$I(r) = -\int_1^{\infty} t^r\, dh(t) = -\int_1^{j} t^r\, dh(t)\,.$$

Let now $p \in (0, +\infty)$. Integrating by parts we obtain

$$I(p) = I(q) + (p-q) \int_1^j t^{p-q-1} \left(- \int_t^j s^q \, dh(s) \right) dt . \tag{1.3}$$

Using assumption 2 we have

$$\int_1^j t^{p-q-1} \left(- \int_t^j s^q \, dh(s) \right) dt \le a \left[\int_1^j t^{p-1} h(t) \, ds + \int_1^j t^{p-1} H(t) \, dt \right] =$$

$$= -\frac{a}{p} h(1) + \frac{a}{p} I(p) - \frac{a}{p} H(1) + \frac{a}{p} \left(- \int_1^\infty t^p \, dH(t) \right)$$

and

$$h(1) \ge \frac{1}{a} I(q) - H(1) .$$

Therefore

$$\int_1^j t^{p-q-1} \left(- \int_t^j s^q \, dh(s) \right) dt \le -\frac{1}{p} I(q) + \frac{a}{p} I(p) + \frac{a}{p} \left(- \int_1^\infty t^p \, dH(t) \right) . \tag{1.4}$$

From (1.3), (1.4) we obtain

$$I(p) \le I(q) - \frac{p-q}{p} I(q) + \frac{a}{p} (p-q) I(p) + a \, \frac{p-q}{p} \int_1^\infty t^p \, dH(t)$$

i.e.

$$\frac{p - a(p-q)}{p} I(p) \le \frac{q}{p} I(q) + a \, \frac{p-q}{p} \int_1^\infty t^p \, dH(t)$$

and the result follows when $h(t) = 0$ for $t \in [j, +\infty)$. In the general case, integrating by parts, since $h(t)$ is nonincreasing, we have

$$j^q h(j) \le \int_j^\sigma t^q \, dh(t) + j^q h(\sigma)$$

and therefore using assumption 1

$$j^q h(j) \le -\int_j^\infty t^q \, dh(t) \qquad j \in (1, +\infty) .$$

Set now

$$h_j(t) = \begin{cases} h(t) & t \in [1, j] \\ 0 & t \in [j, +\infty) \end{cases}$$

h_j is nonincreasing, and

$$-\int_t^\infty s^q \, dh_j(s) \le a[t^q h(t) + t^q H(t)] .$$

Therefore, as in the first part of the proof

$$-\int_1^j t^p \, dh(t) \le -\int_1^j t^p \, dh_j \le \frac{q}{aq-(a-1)q}\left(-\int_1^\infty t^q \, dh(t)\right) +$$

$$+ \frac{a(p-q)}{aq-(a-1)p}\left(-\int_1^\infty t^p \, dH(t)\right)$$

and we obtain the result letting $j \to \infty$. q.e.d.

Now, for convenience, let us suppose that the cube $Q_1(0)$ in Theorem 1.2 is actually the n-cube

$$Q = \left\{x \in R^n : |x_i| < \frac{3}{2} \quad i = 1, \cdots, n\right\}$$

and set

$$C_0 = \left\{x \in R^n : |x_i| < \frac{1}{2} \qquad i = 1, \cdots, n\right\}$$

$$C_k = \{x \in Q : 2^{-k} < \operatorname{dist}(x, \partial Q) \leq 2^{-k+1}\} \qquad k \in \mathbf{N} .$$

Obviously one has

$$Q = \bigcup_k C_k .$$

Finally we need the following subdivision lemma, see [121], which generalizes the well-known argument of subdivision[7)] due to F. Riesz in dimension $n = 1$, known as 'Calderon-Zygmund's subdivision argument', see [42].

LEMMA 1.3. *Let* $g \in L^1(Q)$ $g \geq 0$ *and* $\xi > \fint_Q g\,dx$. *Then for every* $\sigma \geq 3$ *there exists a sequence of* n-*cubes* $\{Q^j_{(k)}\}_{k,j \in \mathbf{N}}$ *with sides parallel to the axes and disjoint interiors such that*

$$(1.5) \qquad \begin{cases} Q^j_{(k)} \subset C_k & \forall j, k \in \mathbf{N} \\[2ex] \sigma^n \cdot 2^{nk} \xi < \displaystyle\fint_{Q^j_{(k)}} g\,dx \leq \sigma^n \cdot 3^n \cdot 2^{nk} \cdot \xi & \\[2ex] g \leq \sigma^n \cdot 2^n \cdot 2^{nk} \cdot \xi \ \text{ a.e. on } C_k \setminus \bigcup_j Q^j_{(k)} & \forall k \in \mathbf{N} . \end{cases}$$

[7)] LEMMA. *Let* Q_0 *be an* n-*cube,* $u \in L^1(Q_0)$ *and* k *be a positive constant such that*

$$\fint_{Q_0} |u|\,dx \leq k .$$

Then there exists a sequence of open disjoint n-*cubes* $Q_i \subset Q_0$ *with sides parallel to the ones of* Q_0 *such that*

i) $|u| \leq k$ a.e. *in* $Q_0 \setminus \bigcup Q_i$

ii) $k \leq |u|_{Q_i} \leq 2^n k \quad \forall i$

iii) $\sum_i |Q_i| \leq k^{-1} \int_{Q_0} |u|\,dx$.

Proof. For each fixed k, we use the Calderon-Zygmund argument, making the first subdivision in 3^n cubes and all the others in 2^n cubes. This way we obtain a sequence of cubes $\{P^i_{(k)}\}$ which satisfy the last two conditions in (1.5). Now we have

$$|P^j_{(k)}| \leq \xi^{-1}\sigma^{-n}\cdot 2^{-nk}\cdot 3^n \fint_Q g\,dx \leq \left(\frac{3}{\sigma}\right)^n\cdot 2^{-nk} \leq 2^{-nk}\,.$$

Therefore either $P^j_{(k)} \cap C_k = \phi$ or $P^j_{(k)} \subset C_k$. Taking for each k the cubes $P^i_{(k)}$ with nonempty intersection with $\mathring{C}_k$ we get the result. q.e.d.

REMARK 1.1. The diameter of the n-cube $Q^j_{(k)}$ in Lemma 1.3 is less or equal to $\frac{3}{\sigma}\sqrt{n}\,2^{-k}$, hence

$$\operatorname{diam} Q^j_{(k)} < \frac{1}{2}\operatorname{dist}(Q^j_{(k)}, \partial Q)$$

provided $\sigma > 6\sqrt{n}$.

Proof of Theorem 1.2. Choose $\sigma > 6\sqrt{n}$ and set

$$a_k = (\sigma^n\cdot 2^{nk})^{1/q}$$

$$G(x) = \begin{cases} \dfrac{g(x)}{\|g\|_{q,Q} + \|f\|_{q,Q}} & \text{if } \theta = 0 \\[2ex] \dfrac{\theta^{1/2} g(x)}{\|g\|_{q,Q} + \|f\|_{q,Q}} & \text{if } \theta > 0 \end{cases} \qquad \mathcal{G}(x) = a_k^{-1} G(x) \text{ in } C_k$$

$$F(x) = \begin{cases} \dfrac{f(x)}{\|g\|_{q,Q} + \|f\|_{q,Q}} & \text{if } \theta = 0 \\[2ex] \dfrac{\theta^{1/2} f(x)}{\|g\|_{q,Q} + \|f\|_{q,Q}} & \text{if } \theta > 0 \end{cases} \qquad J(x) = a_k^{-1} F(x) \text{ in } C_k$$

$$E(h,t) = \{x \in Q : h(x) > t\}\,.$$

The first step consists in proving that

$$(1.6)\qquad \int_{E(\mathcal{G},t)} \mathcal{G}^q\,dx \le a\left[t^{q-1}\int_{E(\mathcal{G},t)} \mathcal{G}\,dx + t^{q-1}\int_{E(\tilde{J},t)} \tilde{J}\,dx\right]$$

for $t \in [1,+\infty)$, where a is a constant which depends on n, q, b, θ and $\tilde{J}$ is given by

$$\tilde{J} = M^{\frac{1}{q}}(J^q)\,.$$

We begin by remarking that assumption (1.1) can be rewritten as

$$(1.1)'\qquad M_{\frac{1}{2}d(x)}(G^q) \le \bar{B}M^q(G+\tilde{F}) + \theta M(G^q) \qquad \tilde{F} = M^{\frac{1}{q}}(F^q)$$

in fact from (b) it follows $M^{\frac{1}{q}}(F^q) \le M(M^{\frac{1}{q}}(F^q))$. Now fix $t \in [1,+\infty)$ and set

$$(1.7)\qquad s = \lambda\cdot t$$

where λ is a constant >1 to be chosen later. Since

$$\fint_Q G^q \le \begin{cases} 1 & \text{if} \quad \theta = 0 \\ \sqrt{\theta} & \text{if} \quad \theta \neq 0 \end{cases}$$

we can employ Lemma 1.3 to obtain a disjoint sequence of n-cubes $Q^j_{(k)}$ such that

$$Q^j_{(k)} \subset C_k \qquad \forall j,k \in \mathbf{N}$$

$$(\alpha_k s)^q < \fint_{Q^j_{(k)}} G^q\,dx \le \sigma^n(\alpha_k s)^q$$

$$G \le \alpha_k s \quad \text{in} \quad C_k \setminus \bigcup_j Q^j_{(k)}$$

that is

$$s^q < \fint_{Q^j_{(k)}} \mathcal{G}^q \, dx \leq \sigma^n s^q$$

$$\mathcal{G} \leq s \quad \text{in} \quad C_k \setminus \cup Q^j_{(k)} .$$

Then

$$|E(\mathcal{G},s) \setminus \bigcup_{j,k} Q^i_{(k)}| = 0$$

and hence

$$\int_{E(\mathcal{G},s)} \mathcal{G}^q \, dx \leq \sum_{j,k} \int_{Q^j_{(k)}} \mathcal{G}^q \, dx \leq \sigma^n s^q \sum_{j,k} |Q^j_{(k)}| . \tag{1.8}$$

Now we want to estimate the right-hand side of (1.8). We begin by eliminating the term $\theta M^q(G^q)$ in (1.1)′. Set $\tilde{Q} = \{x \in Q : (1.1) \text{ is true}\}$, $\tilde{\tilde{Q}} = \bigcup_{j,k} Q^j_{(k)}$, and note that $|\tilde{Q}| = |Q|$. For $x \in Q^j_{(k)}$, since $\operatorname{diam} Q^j_{(k)} < \frac{1}{2} \operatorname{dist}(Q^j_{(k)}, \partial Q)$ we have

$$\alpha^q_k s^q < \fint_{Q^j_{(k)}} G^q \, dx \leq n^{\frac{n}{2}} M_{\frac{1}{2}d(x)}(G^q) \leq n^{\frac{n}{2}} M(G^q)$$

and, since $\sqrt{\theta} < 1$, there exists a ball B around x such that

$$n^{-\frac{n}{2}} \sqrt{\theta}\, \alpha^q_k s^q \leq \sqrt{\theta} M(G^q) < \frac{1}{|B|} \int_B G^q \, dx . \tag{1.9}$$

Therefore $(\theta \neq 0)$

$$|B| \leq \frac{1}{\alpha^q_k} \frac{n^{n/2}}{\sqrt{\theta}} \int_B G^q \, dx \leq \frac{n^{n/2}}{\alpha^q_k}$$

i.e.

$$\text{radius of } B \leq \frac{1}{6} 2^{-k} < \frac{1}{2} \operatorname{dist}(x, \partial Q) .$$

Then from (1.9) it follows that

$$M(G^q)(x) \leq \frac{1}{\sqrt{\theta}} M_{\frac{1}{2}d(x)}(G^q)$$

and hence for $x \in \tilde{Q}$

$$M_{\frac{1}{2}d(x)}(G^q)(x) \leq \frac{\bar{B}}{1-\sqrt{\theta}} M^q(G+\tilde{F})(x) . \tag{1.10}$$

Note that (1.10) is obviously true if $\theta = 0$. Now for $x \in Q^j_{(k)} \cap \tilde{Q}$

$$(\alpha_k s)^q \leq \fint_{Q^j_{(k)}} G^q dx \leq n^{\frac{n}{2}} M_{\frac{1}{2}d(x)}(G^q) \leq \frac{n^{\frac{n}{2}} \bar{B}}{1-\sqrt{\theta}} M^q(G+\tilde{F})$$

so there exists a ball B around x such that

$$(\alpha_k s)^q < \frac{n^{\frac{n}{2}} \bar{B}}{1-\sqrt{\theta}} \left(\fint_B (G+\tilde{F})\, dx \right)^q \tag{1.11}$$

i.e.

$$|B| < \frac{n^{\frac{n}{2}} \bar{B}}{1-\sqrt{\theta}} \frac{1}{(\alpha_k s)^q}$$

from which follows that

$$\text{radius of } B < n^{\frac{1}{2}} \left(\frac{\bar{B}}{1-\sqrt{\theta}} \right)^{\frac{1}{n}} \frac{1}{\sigma \cdot \lambda} \cdot 2^{-k} .$$

Now, choosing a suitable $\lambda = \lambda(n, B, \theta)$, we easily see that B has nonempty intersection at most with C_{k-1}, C_k, C_{k+1}. Then we get from (1.11)

$$s < \left(\frac{n^{\frac{n}{2}}\bar{B}}{1-\sqrt{\theta}}\right)^{\frac{1}{q}} \fint_B (\mathcal{G}+\tilde{J})\,dx$$

taking also into account that $\alpha_{k+1} = 2^{n/q}\alpha_k$.

Now

$$\lambda t\,|B| \le \left(\frac{n^{\frac{n}{2}}\bar{B}}{1-\sqrt{\theta}}\right)^{\frac{1}{q}} \left\{ \int_{B\cap E(\mathcal{G},t)} \mathcal{G}\,dx + \int_{B\cap E(\tilde{J},t)} \tilde{J}\,dx + 2t\,|B| \right\}$$

hence, if moreover

$$\lambda > 2\left(\frac{n^{\frac{n}{2}}\bar{B}}{1-\sqrt{\theta}}\right)^{\frac{1}{q}}$$

we finally get

$$|B| \le \frac{c(n,q,\bar{B},\theta)}{t}\left[\int_{B\cap E(\mathcal{G},t)} \mathcal{G}\,dx + \int_{B\cap E(\tilde{J},t)} \tilde{J}\,dx \right].$$

The family of balls B obviously covers $\tilde{\tilde{Q}}$, therefore, using Lemma 1.1, there exists a numerable disjoint subfamily $\{B_i\}$ such that

$$|\tilde{\tilde{Q}}| \le c(n)\sum_i |B_i| \le \frac{c(n,q,b,\theta)}{t}\left[\int_{E(\mathcal{G},t)} \mathcal{G}\,dx + \int_{E(\tilde{J},t)} \tilde{J}\,dx \right]$$

i.e.

$$\int_{E(\mathcal{G},s)} \mathcal{G}^q\,dx \le c(n,q,b,\theta)\,t^{q-1}\left[\int_{E(\mathcal{G},t)} \mathcal{G}\,dx + \int_{E(\tilde{J},t)} \tilde{J}\,dx \right].$$

On the other hand, obviously

$$\int_{E(\mathcal{G},t)\setminus E(\mathcal{G},s)} \mathcal{G}^q\,dx \le c(n,q,b,\theta)\,t^{q-1} \int_{E(\mathcal{G},t)} \mathcal{G}\,dx$$

so that (1.6) is proved.

The second step is now to derive (1.2) from (1.6). Set

$$h(t) = \int_{E(\mathcal{G},t)} \mathcal{G}\,dx$$

$$H(t) = \int_{E(\tilde{J},t)} \tilde{J}\,dx\,.$$

Since for a nonnegative function u and $r \ge 1$

$$\int_{E(u,t)} u^r\,dx = -\int_t^{+\infty} u^{r-1}\,d\left(\int_{E(u,s)} u\,dx\right)$$

we easily see that hypotheses of Lemma 1.2 are satisfied (with p and q replaced by $p-1$, $q-1$), so that, using also property (c), it follows

$$\int_Q \mathcal{G}^p\,dx \le c\left\{\int_Q \mathcal{G}^q\,dx + \int_Q J^p\,dx\right\}$$

which immediately gives the result. q.e.d.

We close this section by noting explicitly that reverse Hölder inequalities in the same ball are very much stronger than the ones with ball and double ball. The following considerations may partially illustrate the difference.

Let us assume that for $f \geq 0$ the following Hölder reverse inequality holds:

$$\fint_{Q_R(x_0)} f^q\,dx \leq b\left(\fint_{Q_R(x_0)} f\,dx\right)^q \qquad \forall R < R_0 \tag{1.12}$$

then for $E \subset B_R$ we have

$$\int_E f\,dx \leq \left(\int_E f^q\,dx\right)^{\frac{1}{q}} |E|^{1-\frac{1}{q}} \leq \left(\fint_{Q_R} f^q\,dx\right)^{\frac{1}{q}} \left(\frac{|E|}{|Q_R|}\right)^{1-\frac{1}{q}} |Q_R| \leq \tag{1.13}$$

$$\leq b \int_{Q_R} f\,dx \left(\frac{|E|}{|Q|}\right)^{1-\frac{1}{q}}$$

and

Proposition 1.2. *Let us assume (1.12). Then there exists a constant* c_1 *such that for all* $R < R_0$

$$\int_{Q_R} f\,dx \leq c_1 \int_{Q_{R/2}} f\,dx\,.$$

Proof. Let us choose $\alpha < 1$ in such a way that

$$b\left(\frac{|Q_R \setminus Q_{\alpha R}|}{|Q_R|}\right)^{1-\frac{1}{q}} = \beta < 1\,.$$

Note that α is independent on R. Then (1.13) with $E = Q_R \setminus Q_{\alpha R}$ yields

$$\int_{Q_R \setminus Q_{\alpha R}} f\,dx \le \beta \int_{Q_R} f\,dx$$

i.e.

$$\int_{Q_R} f\,dx = \int_{Q_{\alpha R}} f\,dx + \int_{Q_R \setminus Q_{\alpha R}} f\,dx \le \int_{Q_{\alpha R}} f\,dx + \beta \int_{Q_R} f\,dx .$$

Therefore

$$\int_{Q_R} f\,dx \le \frac{1}{1-\beta} \int_{Q_{\alpha R}} f\,dx \le \frac{1}{(1-\beta)^2} \int_{Q_{\alpha^2 R}} f\,dx \cdots$$

and by iteration we get the result. q.e.d.

A simple consequence of Proposition 1.2 is the following

PROPOSITION 1.3. *Assume (1.12) holds. Then* f *cannot have a zero of infinite order at* x_0 *except that it is identically zero.*

Proof. We have

$$\int_{Q_R} f\,dx \le c_1^k \int_{Q_{R\cdot 2^{-k}}} f\,dx = c_1^k (R\cdot 2^{-k})^\gamma \frac{1}{(R\cdot 2^{-k})^\gamma} \int_{B_{R\cdot 2^{-k}}} f\,dx .$$

Choose now γ in such a way that $c_1 \cdot 2^{-\gamma} = 1$, then

$$\int_{Q_R} f\,dx \le R^\gamma \frac{1}{(R \cdot 2^{-k})^\gamma} \int_{B_{R\cdot 2^{-k}}} f\,dx$$

and if

$$\lim_{k\to\infty} \frac{1}{(R\cdot 2^{-k})^{\gamma}} \int_{B_{R\cdot 2^{-k}}} f\,dx = 0$$

it follows that $f \equiv 0$. q.e.d.

Let us remark that the order of zero at x_0 is finite and depends on c_1.

One sees immediately that Proposition 1.3 does not hold if instead of (1.12) we assume that

$$\fint_{Q_R} f^q\,dx \leq b\left(\fint_{Q_{2R}} f\,dx\right)^q \qquad \forall\, R < R_0 .$$

2. L^p-*estimates for solutions of nonlinear elliptic systems*

As we have already seen, we have

THEOREM 2.1. *If* $u \in H^1(\Omega, R^N)$ *is a weak solution to system (0.1), then* ∇u *is locally* p-*integrable, i.e.* $|\nabla u| \in L^p_{loc}(\Omega)$, *for some* $p > 2$ *and for* $B_R \subset B_{2R} \subset \Omega$

$$\left(\fint_{B_R} |\nabla u|^p\,dx\right)^{\frac{1}{p}} \leq c\left(\fint_{B_{2R}} |\nabla u|^2\right)^{\frac{1}{2}} . \tag{2.1}$$

In this section we want to prove analogous L^p-estimates for weak solutions to general nonlinear elliptic systems of the type:

$$-D_\alpha A_i^\alpha(x, u, \nabla u) + B_i(x, u, \nabla u) = 0 \qquad i = 1, \cdots, N . \tag{2.2}$$

As far as we know, the first L^p-estimates for solutions of elliptic equations with nonregular coefficients (i.e. L^∞ coefficients) are due to B. V. Boyarskii [33] [34] who considered first order Beltrami's type elliptic systems in dimension two. These results were then extended to all dimensions by N. G. Meyers [213]. The main tool of their proofs is the Calderon-

Zygmund inequality for singular integrals [42] and the method is ultimately based on the L^p-theory for equations with constant coefficients through a perturbation argument. We shall present a simple variant of it at the end of this section.

This method cannot be extended in order to cover solutions of system (2.2). Here we shall rely on a Caccioppoli type estimate and Proposition 1.1. The results we shall present are essentially due to M. Giaquinta-G. Modica [121] and N. G. Meyers, A. Elcrat [216].[8)]

Finally, let us remark that the exponent p is not very large; in fact, in each dimension, it tends to 2 as the ellipticity becomes 'bad', see example 2.2 below due to N. G. Meyers [213]; compare for a result in the opposite direction [145] [175].

Nonlinear elliptic systems. For the sake of simplicity we shall confine ourselves to considering second order nonlinear elliptic systems satisfying controllable or natural growth conditions of order 2, $k = 2$, compare Section 4 of Chapter IV.

Let us remark explicitly that the method and the result can be carried over to higher order systems with general polynomial growth $k > 1$, with only technical complication, see [121] [216].

Let us begin by considering weak solutions to system (2.2), assuming that controllable growth conditions hold, i.e. conditions (I) of Chapter I, Section 5.

I. *we have*

$$|A_i^\alpha(x, u, \nabla u)| \le \mu_1(|\nabla u| + |u|^{r/2} + f_{i\alpha})$$

$$|B_i(x, u, \nabla u)| \le \mu_2(|\nabla u|^{2\left(1-\frac{1}{r}\right)} + |u|^{r-1} + f_i)$$

8) In [216] there are some gaps.

$$r = \begin{cases} \dfrac{2n}{n-2} & \text{if} \quad n > 2 \\ \text{any exponent} & \text{if} \quad n = 2 \end{cases}$$

$$f_{i\alpha} \in L^2(\Omega), \quad f_i \in L^{\frac{r}{r-1}}(\Omega)$$

here elliptic means

(2.3) $$A_i^\alpha(x,u,p)\,p_\alpha^i \geq \lambda|p|^2 - \mu_2|u|^r - f^2 \quad {}^{9)}$$

with $f \in L^2(\Omega)$; λ *is a positive constant and* μ_1, μ_2 *are nonnegative constants.*

Without loss in generality we can and shall suppose $r > 2$.

Let us remark that conditions I are the minimal ones in order to consider weak solutions in the sense that u belongs to $H^{1,2}(\Omega, R^N)$ and satisfies

(2.4) $$\int_\Omega [A_i^\alpha(x,u,\nabla u)D_\alpha\phi^i + B_i(x,u,\nabla u)\phi^i]dx = 0$$

for all $\phi \in H_0^1(\Omega, R^N)$, compare with Section 5, Chapter I.

We have

THEOREM 2.2. *Suppose that I holds and that* $f, f_{i\alpha} \in L^\sigma(\Omega)$, $f_i \in L^s(\Omega)$ *with* $\sigma > 2$ *and* $s > \frac{r}{r-1}$. *Then there exists an exponent* $p > 2$ *such that, if* $u \in H^{1,2}(\Omega, R^N)$ *is a weak solution to (2.2), then* $u \in H^{1,p}_{loc}(\Omega, R^N)$. *Moreover for* $B_{R/2} \subset B_R \subset \Omega$, *we have*

---

9) Let us remark that the strong ellipticity condition

$$A^\alpha_{ip^j_\beta}\,\xi^i_\alpha\xi^j_\beta \geq \lambda|\xi|^2$$

implies (2.3).

$$\left(\fint_{B_{R/2}} (|u|^r + |\nabla u|^2)^{\frac{p}{2}} dx\right)^{\frac{1}{p}} \le c\left\{\left(\fint_{B_R} (|u|^r + |\nabla u|^2) dx\right)^{\frac{1}{2}} + \right.$$

(2.5)

$$\left. + \left(\fint_{B_R} \Big(f^2 + \sum_{1,\alpha} |f_{i\alpha}|^2\Big)^{\frac{p}{2}} dx\right)^{\frac{1}{p}} + R\left(\fint_{B_R} \sum_i |f_i|^{\frac{r}{r-1}\frac{p}{2}} dx\right)^{\frac{2}{p}\frac{r-1}{r}}\right\}$$

provided $R < R_0$, *where* $c = c(n, \lambda, \mu_1, \mu_2, p)$ *and* R_0 *is a constant depending on* u *for* $n > 2$ *and on* u *only through the* L^2*-norm of the gradient of* u *if* $n = 2$.

Proof. Set $\tilde{f} = (f_i)$, $\tilde{\tilde{f}} = (f_{i\alpha})$. Choose as test function in (2.4) $\phi = (u - u_R)\eta^2$ with $\eta \in C_0^\infty(B_R)$, $0 \le \eta \le 1$, $\eta \equiv 1$ on $B_{R/2}$, $|\nabla\eta| \le c/R$. Then we get

$$\lambda \int |\nabla u|^2 \eta^2 dx \le \mu_2 \int |u|^r \eta^2 + \int f^2\eta^2 + \mu_1 \int |\nabla u|\, |u - u_R|\eta|\nabla\eta| +$$

(2.6)

$$+ \mu_1 \int |u|^{\frac{r}{2}} |u - u_R|\eta\nabla\eta + \int |\tilde{\tilde{f}}|\, |u - u_R|\eta|\nabla\eta| +$$

$$+ \mu_1 \int |\nabla u|^{2\left(1-\frac{1}{r}\right)} |u - u_R|\eta^2 dx + \mu_1 \int |u|^{r-1} |u - u_R|\eta^2 +$$

$$+ \int |\tilde{f}|\, |u - u_R|\eta^2 .$$

Now we have

$$\int |\tilde{f}|\,|u-u_R|\eta^2 dx \leq \Big(\int_{B_R} |u-u_R|^r dx\Big)^{\frac{1}{r}} \Big(\int_{B_R} |\tilde{f}|^{\frac{r}{r-1}} dx\Big)^{1-\frac{1}{r}} \leq$$

$$\leq R^{n\left(\frac{1}{r}-\frac{1}{2}\right)+1} \Big(\int_{B_R} |\nabla u|^2 dx\Big)^{\frac{1}{2}} \Big(\int_{B_R} |\tilde{f}|^{\frac{r}{r-1}} dx\Big)^{1-\frac{1}{r}} \leq$$

$$\leq \varepsilon \int_{B_R} |\nabla u|^2 dx + \frac{c}{\varepsilon} R^{2\left[n\left(\frac{1}{r}-\frac{1}{2}\right)+1\right]} \Big(\int_{B_R} |f|^{\frac{r}{r-1}} dx\Big)^{2\left(1-\frac{1}{r}\right)}$$

$$\int |u|^{r-1} |u-u_R|\eta^2 dx \leq \Big(\int_{B_R} |u-u_R|^r dx\Big)^{\frac{1}{r}} \Big(\int_{B_R} |u|^r dx\Big)^{1-\frac{1}{r}} \leq$$

$$\leq \frac{1}{r} \int_{B_R} |u-u_R|^r dx + \frac{r-1}{r} \int_{B_R} |u|^r dx \leq$$

$$\leq c\, R^{r\left[n\left(\frac{1}{r}-\frac{1}{2}\right)+1\right]} \Big(\int_{B_R} |\nabla u|^2 dx\Big)^{\frac{r}{2}-1} \int_{B_R} |\nabla u|^2 + c \int_{B_R} |u|^r dx$$

$$\int |\nabla u|^{2\left(1-\frac{1}{r}\right)} |u-u_R|\eta^2 \leq \Big(\int_{B_R} |u-u_R|^r dx\Big)^{\frac{1}{r}} \Big(\int_{B_R} |\nabla u|^2 dx\Big)^{1-\frac{1}{r}} \leq$$

$$\leq R^{n\left(\frac{1}{r}-\frac{1}{2}\right)+1} \Big(\int_{B_R} |\nabla u|^2 dx\Big)^{\frac{1}{2}-\frac{1}{r}} \int_{B_R} |\nabla u|^2$$

$$\int_{B_R} |u|^r dx \le c \int_{B_R} |u-u_R|^r + c \int_{B_R} |u_R|^r \le$$

$$\le c R^{r\left[n\left(\frac{1}{r}-\frac{1}{2}\right)+1\right]} \left(\int_{B_R} |\nabla u|^2 dx\right)^{\frac{r}{2}-1} \int_{B_R} |\nabla u|^2 +$$

$$+ c\,|B_R| \left(\fint_{B_R} u\,dx\right)^r \le \cdots + c\,|B_R| \left(\fint_{B_R} |u|^{r\frac{q}{2}} dx\right)^{\frac{2}{q}}$$

for $q > 1$.

Hence from (2.6) it follows for $q > 1$ and $\varepsilon > 0$ (adding to the left-hand side $\int_{B_{R/2}} |u|^r dx$)

$$\int_{B_{R/2}} (|\nabla u|^2 + |u|^r)\,dx \le c\,\frac{1}{R^2} \int_{B_R} |u-u_R|^2 dx +$$

$$+ |B_R| \left(\fint_{B_R} |u|^{r\frac{q}{2}} dx\right)^{\frac{2}{q}} + \int_{B_R} (|f|^2 + |\tilde{\tilde{f}}|^2 + |\tilde{F}|^2)\,dx +$$

$$+ c\left[\varepsilon + R^{r\left[n\left(\frac{1}{r}-\frac{1}{2}\right)+1\right]} \left(\int_{B_R} |\nabla u|^2\right)^{\frac{r}{2}-1} + \right. \tag{2.7}$$

$$\left. + R^{n\left(\frac{1}{r}-\frac{1}{2}\right)+1} \left(\int_{B_R} |\nabla u|^2\right)^{\frac{1}{2}-\frac{1}{r}}\right] \int_{B_R} |\nabla u|^2$$

where

$$\tilde{F} = R^{n\left(\frac{1}{r}-\frac{1}{2}\right)} \left(\int_{B_R} |\tilde{f}|^{\frac{r}{r-1}} dx \right)^{\frac{1}{2}\left(1-\frac{2}{r}\right)} \tilde{f}^{\frac{1}{2}\,\frac{r}{r-1}} .$$

Now note that, because of the absolute continuity theorem of Lebesgue, if $n > 2$ (in fact $\frac{r}{2}-1$, $\frac{1}{2}-\frac{1}{r} > 0$) and simply since $n\left(\frac{1}{r}-\frac{1}{2}\right)+1 > 0$ if $n = 2$, the coefficient of $\int_{B_R} |\nabla u|^2$ on the right-hand side of (2.7) goes to $c \cdot \varepsilon$ when $R \to 0$.

Therefore, dividing by R^n, setting

$$q = \frac{2n}{n+2}$$

using the Sobolev-Poincaré inequality and choosing ε suitably, we get for R less than some R_0

$$\begin{aligned} \fint_{B_{R/2}} (|u|^r + |\nabla u|^2)\,dx \le c \Bigg\{ & \left(\fint_{B_R} (|\nabla u|^q + |u|^{r\frac{q}{2}})\,dx \right)^{\frac{2}{q}} + \\ & + \fint_{B_R} (|f|^2 + |\tilde{\tilde{f}}|^2 + |\tilde{F}|^2)\,dx \Bigg\} + \frac{1}{2} \fint_{B_R} |\nabla u|^2 dx \end{aligned} \tag{2.8}$$

which, through Proposition 1.1, obviously gives the result if $\tilde{F} \equiv 0$.

If $\tilde{F}$ is not zero we work as follows. Fix a cube Q_{R_1}; for each $x \in Q_{R_1}$, $R_1 < R_0$, and for each $R < \frac{1}{2}\operatorname{dist}(x, \partial Q_{R_1})$ (2.8) holds with $\tilde{F}$ given this time by

$$\tilde{F} = R_1^{\left[n\left(\frac{1}{r}-\frac{1}{2}\right)+1\right]} \left(\int_{Q_{R_1}} |\tilde{f}|^{\frac{r}{r-1}} dx \right)^{\frac{1}{2}\left(1-\frac{2}{r}\right)} \tilde{f}^{\frac{1}{2}\,\frac{r}{r-1}}$$

and now the result follows as before, for the arbitrarity of the cube Q_{R_1} and with a simple use of Hölder inequality. q.e.d.

An application of Sobolev theorem gives immediately

COROLLARY 2.1. *Under the assumptions of Theorem 2.2, if the dimension* n *is two, then* u *is locally Hölder-continuous.*

Now we want to consider weak solutions to system (2.2) under natural growth conditions, and we shall distinguish two cases:

II. *we have*

$$|A_i^\alpha(x,u,\nabla u)| \le \mu_1(|\nabla u| + f_{i\alpha})$$

$$|B_i(x,u,\nabla u)| \le \mu\ (|\nabla u|^{2-\varepsilon} + f_i) \qquad \varepsilon > 0$$

$$f_{i\alpha} \in L^2(\Omega),\ \ f_i \in L^1(\Omega)$$

and elliptic means

$$A_i^\alpha(x,u,p)\,p_\alpha^i \ge \lambda|p|^2 - \mu_1 f^2$$

with $f \in L^2(\Omega)$; λ, μ, μ_1 *are allowed to depend on* u.

or

III. *the same as in II except that*

$$|B_i(x,u,\nabla u)| \le \mu(|\nabla u|^2 + f_i)\ .$$

Let us recall, compare with Section 5, Chapter I, that now 'weak solution' means that $u \in H^1 \cap L^\infty(\Omega, R^N)$ and (2.4) holds for all $\phi \in H_0^1 \cap L^\infty(\Omega, R^N)$.

Then we have

THEOREM 2.3. *Suppose that II is fulfilled and that* $f, f_{i\alpha} \in L^\sigma(\Omega)$, $\sigma > 2$, $f_i \in L^s(\Omega)$, $s > 1$. *Then there exists an exponent* $p > 2$ *such that, if* $u \in H^{1,2} \cap L^\infty(\Omega, R^N)$ *is a weak solution of system (2.2), then* $u \in H^{1,p}_{loc}(\Omega, R^N)$.

If III holds and $f, f_{i\alpha} \in L^{\sigma}(\Omega)$, $\sigma > 2$, $f_i \in L^s(\Omega)$, $s > 1$, *then there exists an exponent* $p > 2$ *such that, if* $u \in H^{1,2} \cap L^{\infty}(\Omega, R^N)$, $|u| \leq M$, *is a weak solution to system (2.2) with*

$$2\mu M < \lambda \tag{2.9}$$

then $u \in H^{1,p}_{loc}(\Omega, R^N)$.

In either case for $B_{R/2} \subset B_R \subset \Omega$ *we have*

$$\left(\fint_{B_{R/2}} |\nabla u|^p dx\right)^{\frac{1}{p}} \leq c\left\{\left(\fint_{B_R} |\nabla u|^2 dx\right)^{\frac{1}{2}} + \left(\fint_{B_R}\left(|f|^2 + \sum |f_{i\alpha}|^2 + \sum |f_i|\right)^{\frac{p}{2}} dx\right)^{\frac{1}{p}}\right\}$$

provided $R < R_0$ *with* R_0 *depending on* M *and the* L^2*-norm of* ∇u, *where* $c = c(u, \mu, \lambda, \mu_1)$.

Proof. Putting as before $\phi = (u - u_R)\eta^2$ in (2.4) we get

$$\lambda \int |\nabla u|^2 \eta^2 dx \leq \mu_1 \int f^2 \eta^2 + 2\mu_1 \int |\nabla u|\, |u - u_R| \eta |\nabla \eta| +$$

$$+ 2\mu_1 \int |(f_{i\alpha})|\, |u - u_R| \eta |\nabla \eta| + \mu \int |(f_i)|\, |u - u_R| \eta^2 +$$

$$+ \mu \int |\nabla u|^{2-\varepsilon} |u - u_R| \eta^2 \quad \left(\text{or } + \mu \int |\nabla u|^2 |u - u_R| \eta^2\right).$$

Noting now that under assumption III and (2.9)

$$\mu \cdot \int |\nabla u|^2\, |u - u_R| \eta^2 < \lambda \int |\nabla u|^2 \eta^2$$

while under assumption II

$$\mu \int |\nabla u|^{2-\varepsilon} |u - u_R| \le c\Big(\int_{B_R} |\nabla u|^2 dx\Big)^{\frac{2-\varepsilon}{2}} \Big(\int_{B_R} |u - u_R|^{\frac{2}{\varepsilon}}\Big)^{\frac{\varepsilon}{2}} \le$$

$$\le c(M) R^{\varepsilon} \int_{B_R} |\nabla u|^2 dx$$

we get the result as in the proof of Theorem 2.2. q.e.d.

It is worth remarking that a smallness condition like (2.9) is necessary for systems, as shown by the following modification of Frehse's example, example 3.7, Chapter II, due to S. Hildebrandt, K.-O. Widman [164]:

EXAMPLE 2.1. For $n = N = 2$, $u = (u^1, u^2) = (\sin(\sigma \log\log |x|^{-1}),$ $\cos(\sigma \log\log |x|^{-1}))$ is a weak solution to

$$-\Delta u = f(u, \nabla u)$$

where

$$f(u, p) = \Big(u^1 + \frac{u^2}{\sigma},\ u^2 - \frac{u^1}{\sigma}\Big)|p|^2 .$$

Here we have $M = \sup |u| = 1$, $\mu = 1 + \sigma^{-2}$, $\lambda = 1$. And obviously $u \notin H^{1,p}$ for $p > 2$. Therefore we see that for $\mu M > \lambda$ no L^p-estimate can hold.[10)]

Assumption (2.9) can be instead eliminated in the case of equations, $N = 1$. In fact we have

PROPOSITION 2.1. *Suppose that* $N = 1$, *III holds with* f, $f_{i\alpha} \in L^{\sigma}(\Omega)$, $\sigma > 2$, $f_i \in L^s(\Omega)$, $s > 1$. *Then the conclusions of Theorem 2.3 are true.*

---

10) The question whether instead of (2.9) we may only assume $\mu M < \lambda$ or not is open.

Proof. Putting $\phi = (u - u_R)e^{t|u-u_R|^2}\eta^2$, η as before, in[11])

$$\int_\Omega [A^\alpha(x,u,\nabla u)D_\alpha \phi + B(x,u,\nabla u)\phi]\,dx = 0 \qquad \forall \phi \in H_0^1 \cap L^\infty(\Omega)$$

since for scalar functions (it is not true for vector valued functions)

$$4|u-u_R|^2|\nabla u|^2 = |\nabla|u-u_R|^2|^2$$

$$2|u-u_R|\ |\nabla u|^2 = |\nabla u|\ |\nabla|u-u_R|^2|$$

we get

$$\int |\nabla u|^2 e^{t|u-u_R|^2}\eta^2\,dx + \frac{t}{2}\int |\nabla|u-u_R|^2|e^{t|u-u_R|^2}\eta^2\,dx \le$$

$$\le \text{const}\left\{\int |\nabla u|\ |u-u_R|\,e^{t|u-u_R|^2}\eta|\nabla\eta| + \right.$$

$$+ \int |\nabla u|\ |\nabla|u-u_R|^2|\,e^{t|u-u_R|^2}\eta^2 +$$

$$\left. + \int_{B_R} (|(f_\alpha)|^2 + |f|)\,e^{t|u-u_R|^2}\right\}$$

and choosing t sufficiently large

---

11) Compare with [191] [164].

$$\int |\nabla u|^2 e^{t|u-u_R|^2} \eta^2 \leq \text{const} \left\{ \int |u-u_R|^2 e^{t|u-u_R|^2} |\nabla\eta|^2 + \right.$$

$$\left. + \int_{B_R} (|(f_a)|^2 + |f|) e^{t|u-u_R|^2} \right\}.$$

To conclude it is sufficient now to proceed as before, taking into account that $e^{t|u-u_R|^2}$ is estimated by constants from above and below. q.e.d.

From Sobolev's theorem we have

COROLLARY 2.2. *Under the assumptions of Theorem 2.3 or of Proposition 2.1, if* $n = 2$, *then* u *is locally Hölder-continuous.*

If we had considered higher order systems, say of order $2m$, with polynomial growth $k > 1$, compare with Section 4, Chapter IV, the results of Corollaries 2.1, 2.2 would sound as: *if* $mk = n$ *then* u *is Hölder-continuous.* In this general setting, this has been proved by Morrey [231] in case of functionals (compare also with Section 3), and with a different method (the 'hole filling technique' [300]) by K. O. Widman [300] and T. G. Todorov [293]. These authors show that $|\nabla u|$ belongs to $L^{2,\lambda}$ (compare Section 1, Chapter III) which is a weaker result with respect to the p-integrability, but enough for deducing the Hölder continuity via the Dirichlet growth theorem, see Chapter III.

We would like to remark that, in the borderline case $mk = n$ Frehse [90] had proved the boundedness of solutions and I. V. Skrypnik [270] [271] [272] the continuity of solutions.

REMARK 2.1. If

$$A_i^\alpha(x, u, \nabla u) = \sum_{j=1}^{N} \sum_{\beta=1}^{n} A_{ij}^{\alpha\beta} D_\beta u^j + a_i^\alpha(x, u)$$

with

$$A_{ij}^{\alpha\beta} = A_{ij}^{\alpha\beta}(x) \in C^0(\Omega)$$

and all the assumptions on the 'lower order terms' remain unchanged, the results of this section are true under the Legendre-Hadamard condition

$$A_{ij}^{\alpha\beta} \xi_\alpha \xi_\beta \lambda^i \lambda^j \geq \nu |\xi|^2 |\lambda|^2 \qquad \forall \xi, \lambda ; \nu > 0 .$$

To see this, one only has to work with the lower order terms as before; while on the leading term one has to localize and use Fourier transform as in the proof of Gärding's inequality [102][2][239], compare with Chapter III.

REMARK 2.2. Under natural growths II, for instance in the case

$$(2.10) \qquad -\Delta u = |\nabla u|^\gamma \qquad 1 + \tfrac{2}{n} < \gamma < 2 , \quad n \geq 3$$

we have considered weak solutions $u \in H^1 \cap L^\infty$ satisfying

$$(2.11) \qquad \int \nabla u \, \nabla \phi \, dx = \int |\nabla u|^\gamma \phi$$

for all $\phi \in H_0^1 \cap L^\infty$. Actually, the notion of solution can be weakened, in fact (2.11) has a meaning for all $\phi \in H_0^1 \cap L^{2/2-\gamma}$. Therefore one could be led to consider as solutions to system (2.10) functions $u \in H^1 \cap L^{2/2-\gamma}$ satisfying (2.11) for all $\phi \in H_0^1 \cap L^{2/2-\gamma}$.

But Ladyzhenskaya and Ural'tseva have shown, for equations, that this is not a good starting point for weak solutions, compare the introduction and Section 7, Chapter 4 of [191]; in fact, if we want to prove Hölder continuity of weak solutions, it is necessary (and sufficient for equations) to have

$$(2.12) \qquad u \in H^{1,2} \cap L^{n \frac{\gamma-1}{2-\gamma}} .$$

Actually in this situation an L^p-estimate holds, too. We confine ourselves to showing that in the simple case of equation (2.11). Inserting $\phi = (u-u_R)\eta^2$ in (2.11), η^2 the standard cut off function, we get

$$\int_{B_{R/2}} |\nabla u|^2 dx \le \frac{c}{R^2} \int_{B_R} |u-u_R|^2 + \int_{B_R} |\nabla u|^\gamma |u-u_R|$$

and since

$$\int_{B_R} |\nabla u|^\gamma |u-u_R| dx \le \left(\int_{B_R} |\nabla u|^2\right)^{\frac{\gamma}{2}} \left(\int_{B_R} |u-u_R|^{\frac{2}{2-\gamma}} dx\right)^{1-\frac{\gamma}{2}} \le$$

$$\le \left(\int_{B_R} |\nabla u|^2\right)^{\frac{\gamma}{2}} \left[\left(\int_{B_R} |u-u_R|^{2^*} dx\right)^{\frac{2}{2^*}} \left(\int_{B_R} |u-u_R|^{n\frac{\gamma-1}{2-\gamma}} dx\right)^{1-\frac{2}{2^*}}\right]^{1-\frac{\gamma}{2}} \le$$

$$\le \int_{B_R} |\nabla u|^2 \cdot \left[\int_{B_R} |u|^{n\frac{\gamma-1}{2-\gamma}} dx\right]^{\frac{2}{n}\left(1-\frac{\gamma}{2}\right)}$$

the result follows in the standard way.

Finally we would like to point out that the methods and results of this section have been extended to studying certain variational inequalities, see L. Boccardo [29], M. Giaquinta [108] [109], and to studying systems of the type of the stationary Navier-Stokes system, see M. Giaquinta, G. Modica [123].

A system in variation. Let us suppose that controllable growth conditions hold and that A_i^α, B_i are smooth. Then we can derive the system in variation and if the polynomial growth is $k = 2$, we get a fourth order system of the type considered in the previous subsection, compare with

Section 1, Chapter II; therefore we obtain an L^p-estimate also for the second derivatives of the weak solution to system (2.2), see [121] [122].

This is not true anymore if we are considering polynomial growth $k > 1$; then the system in variation is of the form: (recall that now v stands for ∇u):

$$(2.13) \qquad \int_\Omega V^{k-2}[A_{ij}^{\alpha\beta}(x,v)D_\beta v^j + b_i^\alpha(x,v)]D_\alpha \phi^i = 0 \qquad \forall \phi \in C_0^\infty(\Omega, R^N)$$

compare with Section 4, Chapter IV, and it is not of the type considered up to now in this section. Still it would be interesting to have an L^p-estimate especially in order to obtain an estimate of the Hausdorff dimension of the singular set, compare with Theorems 4.3, 4.3′ of Chapter IV.

Let us consider weak solutions to system (2.13) where, we recall

$$V = V(u) = (1+|u|^2)^{\frac{1}{2}}$$

$$|A_{ij}^{\alpha\beta}| \le L$$

$$A_{ij}^{\alpha\beta}\xi_\alpha^i\xi_\beta^j \ge |\xi|^2$$

and, for the sake of simplicity, assume $b_i^\alpha \equiv 0$. Then we have

THEOREM 2.4. *There exists an exponent* $p > 2$ *such that* $V^{k/2} \in H^{1,p}_{loc}(\Omega)$. *Here we are assuming the polynomial growth* $k > 2$.

Proof.[12] Inserting in (2.13) $\phi = (u-\lambda)\eta^2$, η being a standard cut-off function, we obtain

---

[12] This proof is due to E. Giusti (private communication, beginning of 1979).

$$\int_{B_{R/2}} V^{k-2}|\nabla u|^2 dx \le \frac{c}{R^2}\int_{B_R} V^{k-2}|u-\lambda|^2 dx$$

Now it is not difficult to verify that

$$\frac{V^{k-2}(u)\,|u-\lambda|^2}{|V^{\frac{k-2}{2}}(u)u - V^{\frac{k-2}{2}}(\lambda)\lambda|^2} \le \text{absolute constant}$$

therefore, choosing λ in such a way that $V^{\frac{k-2}{2}}(\lambda)\lambda = (V^{\frac{k-2}{2}}(u)u)_R$ and and applying Sobolev-Poincaré inequality we obtain

$$\int_{B_{R/2}} |\nabla V^{\frac{k}{2}}|^2 dx \le \frac{c}{R^2}\left(\int_{B_R} |\nabla V^{\frac{k}{2}}|^q\right)^{\frac{2}{q}} \qquad q = \frac{2n}{n+2}$$

and the result follows as in the theorems of the previous subsection.

q.e.d.

The result and the proof of Theorem 2.4 do not extend straightforwardly to solutions of the higher order systems (4.5) in Chapter IV, nor even to solutions of second order systems with $1 < k < 2$.

Boundary estimates. Since, roughly speaking, Caccioppoli inequalities hold up to the boundary, the method for obtaining higher integrability described above can be carried over up to the boundary. We shall not do that for general boundary value problems and we shall confine ourselves to the Dirichlet problem, moreover for the sake of simplicity, we shall restrict ourselves to considering weak solutions to the Dirichlet problem for the Laplace operator

$$\begin{cases} \displaystyle\int_\Omega \nabla u \nabla \phi \, dx = 0 & \forall \phi \in H_0^1(\Omega) \\ u - v \in H_0^1(\Omega) \end{cases}$$

assuming that $\partial\Omega$ be smooth and that the boundary value $v \in H^{1,s}(\Omega)$ for some $s > 2$.

Let Q_{R_0} be an n-cube in R^n with $Q_{R_0} \cap \Omega \neq \phi$. For $x \in Q_{R_0}$ and $R < \frac{1}{2}\operatorname{dist}(x, \partial Q_{R_0})$ we have three possibilities:

1. $Q_{\frac{3}{2}R}(x) \cap \Omega = \phi$
2. $Q_{\frac{3}{2}R}(x) \cap [Q_{R_0} \setminus \Omega] = \phi$
3. $Q_{\frac{3}{2}R}(x) \cap \Omega \neq \phi \qquad Q_{\frac{3}{2}R}(x) \cap (Q_{R_0} \setminus \Omega) \neq \phi$.

In case 2, as we have seen, we have

$$\fint_{Q_R(x)} |\nabla u|^2 dx \leq c \left(\fint_{Q_{\frac{3}{2}R}(x)} |\nabla u|^q dx \right)^{\frac{2}{q}} \qquad q = \frac{2n}{n+2}.$$

In case 3 we have, since for $\eta \in C_0^1(B_{\frac{3}{2}R}(x))$ $\eta \equiv 1$ on $Q_R(x)$ $(u-v)\eta \in H_0^1(\Omega)$,

$$\int_{Q_R} |\nabla u|^2 \leq \frac{c}{R^2} \int_{Q_{\frac{3}{2}R} \cap \Omega} |u-v|^2 + c \int_{Q_{\frac{3}{2}R} \cap \Omega} |\nabla v|^2 \leq$$

$$\leq \text{const} \left\{ \frac{1}{R^2} \int_{Q_{2R} \cap \Omega} |u-v|^2 + \int_{Q_{2R} \cap \Omega} |\nabla v|^2 \right\}.$$

And, since $\partial\Omega$ is smooth (and Ω bounded), we have $\text{meas}\,(Q_{2R}(x)\setminus\Omega) \geq \gamma\ \text{meas}\ Q_{2R}$ for some $\gamma > 0$; hence, extending u, v zero outside of $Q_{2R_0} \cap \Omega$, we get[13)]

$$\fint_{Q_R} |\nabla u|^2 \leq \text{const}\left\{\left(\fint_{Q_{2R}\cap\Omega} |\nabla u - \nabla v|^q dx\right)^{\frac{2}{q}} + \fint_{Q_{2R}\cap\Omega} |\nabla v|^2\right\}.$$

In conclusion if we set

$$g(x) = \begin{cases} |\nabla u|^q & \text{for} \quad x \in \Omega \cap Q_{R_0} \\ 0 & \text{for} \quad x \in Q_{R_0} \setminus \Omega \end{cases}$$

$$f(x) = \begin{cases} |\nabla v|^q & \text{for} \quad x \in \Omega \cap Q_{R_0} \\ 0 & \text{for} \quad x \in Q_{R_0} \setminus \Omega \end{cases}$$

we obtain $\forall\, x \in Q_{R_0}$ and $R < \frac{1}{2}\,\text{dist}\,(x, \partial Q_{R_0})$

$$\fint_{Q_R(x)} g^{\frac{2}{q}} dx \leq c\left\{\left(\fint_{Q_{2R}(x)} g\right)^{\frac{2}{q}} + \int_{Q_{2R}(x)} |\nabla v|^{\frac{2}{q}} dx\right\}$$

which implies, through Proposition 1.1,

---

13) Here we use the following Sobolev-Poincaré theorem

PROPOSITION. *Let* $u \in H^{1,p}(Q_R)$, $p > 1$. *Suppose that* $\mathcal{L}^n\{x \in Q_R : u(x) = 0\} \geq \mu \mathcal{L}^n Q_R$, $\mu > 0$, *then*

$$\left(\int_{Q_R} |u|^{p^*} dx\right)^{\frac{1}{p^*}} \leq c\left(\int_{Q_R} |\nabla u|^p\right)^{\frac{1}{p}}$$

where c *is a constant independent of* R *and* p^* *is the Sobolev exponent of* p.

$$\left(\fint_{Q_{R_{0/2}}\cap\Omega} |\nabla u|^p\,dx\right)^{\frac{1}{p}} \le c\left\{\left(\fint_{Q_{R_0}\cap\Omega} |\nabla u|^2\,dx\right)^{\frac{1}{2}} + \left(\fint_{Q_{R_0}\cap\Omega} |\nabla v|^p\right)^{\frac{1}{p}}\right\}$$

for some $p > 2$.

The perturbation argument. Let us describe how to obtain an L^p-estimate for solutions of linear systems with L^∞ coefficients relying on the L^p theory for systems with constant coefficients (the Laplace operator) and a perturbation argument, see [213] [240] [50].

First let us consider the weak solution $u \in H^1_0(\Omega, R^N)$ [14)] of

$$-D_\alpha[A^{\alpha\beta}_{ij}(x)D_\beta u^j] + D_\alpha f^\alpha_i - f_i = 0 \qquad i = 1, \cdots, N$$

where $A^{\alpha\beta}_{ij} \in L^\infty(\Omega)$

$$|A^{\alpha\beta}_{ij}| \le L$$

$$A^{\alpha\beta}_{ij}\xi^i_\alpha\xi^j_\beta \ge \nu|\xi|^2 \qquad \forall\xi;\ \nu > 0$$

and to simplify assume that $A^{\alpha\beta}_{ij} = A^{\beta\alpha}_{ji}$.

Then we have

THEOREM 2.5. *There exists a number* p_0 *such that for* $2 < p < p_0$ $\nabla u \in L^p$ *and*

$$\int_\Omega \sum_{i,\alpha} |D_\alpha u^i|^p\,dx \le c\left\{\int_\Omega \sum |f^\alpha_i|^p + \int_\Omega |f|^2\,dx\right\} \qquad \forall p,\ 2 \le p < p_0 .$$

Proof. Let $v \in H^1_0(\Omega, R^N)$ be the solution to

$$-\Delta v^i + D_\alpha f^\alpha_i = 0 \qquad i = 1, \cdots, N .$$

---

14) We assume Ω smooth.

We have

$$\int_\Omega |\nabla v|^2 dx \le \int_\Omega |\tilde{f}|^2 dx \qquad \tilde{f} = (f_i^\alpha)$$

and from the L^p-theory for the Laplace operator

$$\int_\Omega \sum_{\alpha,i} |D_\alpha v^i|^p\, dx \le c(p) \int_\Omega \sum |f_i^\alpha|^p\, dx\ .$$

For $r > 2$ fixed, Riesz-Thorin interpolation theorem,[15] see e.g. [286], tells us that

$$c(p) = [c(r)]^{\frac{r(p-2)}{p(r-2)}} \qquad 2 < p < r$$

hence $c(p)$ is a continuous and nondecreasing function of p in $[2,r]$.

Let $w \in H_0^1(\Omega, R^N)$ be the solution to

$$-\Delta w^i - f_i = 0 \qquad i = 1, \cdots, N$$

we have $|\nabla^2 w| \in L^2(\Omega)$, in particular $|\nabla w| \in L^{2^*}(\Omega)$, and

$$\Sigma \|D^\alpha w^i\|_{L^{2^*}(\Omega)} \le c\|f\|_{L^2(\Omega)}\ .$$

---

15) RIESZ-THORIN THEOREM. *Let* $u \to Tu$ *be linear and* (p_j, q_j)*-strong, i.e.*

$$\|Tu\|_{L^{q_j}(\Omega)} \le k_j \|u\|_{L^{p_j}(\Omega)} \qquad j = 1, 2\ .$$

Then, for all $t \in [0,1]$, T *maps* $L^p(\Omega)$ *into* $L^p(\Omega)$ *where*

$$\frac{1}{p} = \frac{1-t}{p_1} + \frac{t}{p_2} \qquad \frac{1}{q} = \frac{1-t}{q_1} + \frac{t}{q_2}$$

and moreover

$$\|Tu\|_{L^q(\Omega)} \le k_1^{1-t} k_2^t \|u\|_{L^p(\Omega)}\ .$$

Then $u \in H_0^1(\Omega, R^N)$ satisfies

$$\int_\Omega A_{ij}^{\alpha\beta} D_\beta u^j D_\alpha \phi^i \, dx = \int_\Omega h_i^\alpha D_\alpha \phi^i \qquad \forall \phi \in H_0^1(\Omega, R^N)$$

where

$$h_i^\alpha = f_i^\alpha + D_\alpha w^i .$$

Let us now consider the linear transformation

$$T : H_0^{1,p}(\Omega, R^N) \to H_0^{1,p}(\Omega, R^N)$$

defined as: Tu is the unique solution to

$$\int_\Omega \nabla v \cdot \nabla \phi = \int_\Omega \left(\delta_{ij}\delta_{\alpha\beta} - \frac{A_{ij}^{\alpha\beta}}{L}\right) D_\beta u^j D_\alpha \phi^i + \frac{1}{L}\int_\Omega h_i^\alpha D_\alpha \phi^i \quad \forall \phi \in H_0^1(\Omega, R^N).$$

Remarking that

$$\left|\left(\delta_{ij}\delta_{\alpha\beta} - \frac{A_{ij}^{\alpha\beta}}{L}\right)\xi_\alpha^i \eta_\beta^j\right| \le$$

$$\le \left\{\left(\delta_{ij}\delta_{\alpha\beta} - \frac{A_{ij}^{\alpha\beta}}{L}\right)\xi_\alpha^i \xi_\beta^j\right\}^{\frac{1}{2}} \left\{\left(\delta_{ij}\delta_{\alpha\beta} - \frac{A_{ij}^{\alpha\beta}}{L}\right)\eta_\alpha^i \eta_\beta^j\right\}^{\frac{1}{2}}$$

we get

$$\left(\int_\Omega \sum_{\alpha,i}\left[\sum_{\beta,j}\left(\delta_{\alpha\beta}\delta_{ij} - \frac{A_{ij}^{\alpha\beta}}{L}\right) D_\beta u^j\right]^p dx\right)^{\frac{1}{p}} = \sup_{\sum \|g_i^\alpha\|_{L^{p'}(\Omega)} = 1} \int_\Omega \left(\delta_{ij}\delta_{\alpha\beta} - \frac{A_{ij}^{\alpha\beta}}{L}\right) D_\beta u^j g_i^\alpha$$

$$\le \left(1 - \frac{\nu}{L}\right)\left(\int_\Omega \sum_{\alpha,i} |D_\alpha u^i|^p \, dx\right)^{\frac{1}{p}} .$$

Therefore for $v = Tu$

$$\int_\Omega \sum |D_\alpha v^i|^p dx \le c(p)\left\{\int_\Omega |h_i^\alpha|^p dx + \left(1-\frac{\nu}{L}\right)\int_\Omega \sum_{\alpha,i} |D_\alpha u^i|^p dx\right\}$$

and for $v_1 = Tu_1$ and $v_2 = Tu_2$

$$\int_\Omega \sum |D_\alpha(v_1^i - v_2^i)|^p dx \le c(r)^{\frac{r(p-2)}{p(r-2)}}\left(1-\frac{\nu}{L}\right)\int_\Omega \sum |D_\alpha(u_1^i - u_2^i)|^p dx$$

and we see that, for p smaller than some $p_0 > 2$, T is a contraction, hence there exists a unique fixed point u_0, which obviously coincides with the solution u. q.e.d.

The local estimate can now be obtained by remarking that ηu, with $\eta \in C_0^\infty(B_R)$, satisfies

$$-D(A\,D(\eta u)) = -D\eta\, A\, Du + D(A u\, D\eta)\ .$$

EXAMPLE 2.2 (N. G. Meyers [213]). Let (x, y) be the coordinates in the plane. Consider the equation

$$(au_x + bu_y)_x + (bu_x + cu_y)_y = 0 \tag{2.14}$$

where

$$a = 1 - (1-\mu^2)\frac{y^2}{x^2+y^2}$$

$$b = (1-\mu^2)\frac{xy}{x^2+y^2}$$

$$c = 1 - (1-\mu^2)\frac{x^2}{x^2+y^2}$$

and μ is a fixed constant, $0 < \mu < 1$. Then we have: the eigenvalue of the coefficient matrix are μ^2 and 1, the function

$$u(x,y) = (x^2+y^2)^{\frac{\mu-1}{2}} x$$

is a solution to (2.14) and $|\nabla u| \in L^p_{loc}$ for $p < \frac{2}{1-\mu}$, but

$$\int_{B_1} |\nabla u|^{\frac{2}{1-\mu}} dx\,dy = +\infty .$$

Therefore we have

$$2 < p_0 < \frac{2}{1-\mu}$$

and we see that for $\mu \to 0$, $p_0 \to 2$. Analogous examples can be constructed in each dimension. Note that for $\mu \to 1$, $p_0 \to \infty$, compare with [145] [175].

3. *An* L^p*-estimate for minimum points of nondifferentiable functionals*

Let us consider the multiple integral

$$J[u] = \int_\Omega F(x, u, \nabla u)\,dx \tag{3.1}$$

where we assume

i) $F(x,u,p)$ *is measurable in* x *for all* $(u,p) \in R^N \times R^{nN}$ *and continuous in* (u,p) *for a.e.* $x \in \Omega$

ii) *there exist two positive constants* μ, Λ *and a nonnegative constant* K *such that*

$$\mu|p|^m - K \le F(x,u,p) \le \Lambda|p|^m + K$$

for all $(x,u,p) \in \Omega \times R^N \times R^{nN}$, *where* m *is a real number* > 1.

In this section we want to show that the gradient of minimum points of $J[u]$ in (3.1) under the assumptions i) ii) is higher integrable. Note that it is not assumed that $F(x, u, p)$ be convex with respect to p. More precisely we have

THEOREM 3.1. *Let* $u \in H^{1,m}(\Omega, R^N)$ *be a minimum point for* $J[u]$ [16]) *and let i) ii) hold. Then there exists an exponent* $q > m$ *such that* $u \in H^{1,q}_{loc}(\Omega, R^N)$; *moreover for all* $x_0 \in \Omega$ *and* $R < \operatorname{dist}(x_0, \partial\Omega)$ *the following estimate holds*:

$$(3.2) \qquad \left(\fint_{B_{R/2}(x_0)} (1 + |\nabla u|)^q \, dx \right)^{\frac{1}{q}} \le c \left(\fint_{B_R(x_0)} (1 + |\nabla u|)^m \, dx \right)^{\frac{1}{m}}$$

where c *is a constant independent of* u.

This result is due to M. Giaquinta, E. Giusti [114].

Before proving it, let us make a few remarks. In Section 2 we have proved various L^p-estimates, but none of those results applies to the case considered in Theorem 3.1. In fact first of all we are not assuming any differentiability on F and no ellipticity, i.e. convexity of F in p. But even if we assumed that F be smooth and strictly convex in p, we would have to require growth conditions on F_u, F_p; and if natural growth conditions held, then we could show that (3.2) holds only if u is bounded and F_u satisfies a smallness condition, compare with Section 2.

---

16)This means that

$$J[u] \le J[v]$$

for all $v \in H^{1,m}(\Omega, R^N)$ with $u - v \in H_0^{1,m}(\Omega, R^N)$, or more weakly (if Ω is unbounded)

$$J_{\tilde{\Omega}}[u] \le J_{\tilde{\Omega}}[v]$$

for all $\tilde{\Omega} \subset\subset \tilde{\Omega}$ and $v \in H^{1,m}(\tilde{\Omega}, R^N)$ with $u - v \in H_0^{1,m}(\tilde{\Omega}, R^N)$.

We are allowed to use the results of Section 2 only if

1) F is smooth, $F(x,u,p) = F(x,p)$ (i.e. there is no explicit dependence on u) and F satisfies natural growth conditions or

2) F is smooth, satisfies natural growths, $N = 1$ and u is bounded.

Let us now come to the proof.

Proof of Theorem 3.1. Fix $x_0 \in \Omega$ and $t < s < R$ in such a way that $B_R(x_0) \subset \Omega$ and choose $\eta \in C_0^\infty(B_s)$ $0 \le \eta \le 1$ with $\eta \equiv 1$ on B_t and $|\nabla\eta| \le \frac{c_1}{s-t}$. The function

$$v = u + \eta(u - u_R)$$

is a competing function, therefore

$$J[u] \le J[v]$$

which implies

$$\mu \int_{B_s} |\nabla u|^m dx - K|B_s| \le c(m)\Lambda \int_{B_s} (1-\eta)^m |\nabla u|^m +$$

$$+ c(m)\Lambda \int_{B_s} |\nabla\eta|^m |u - u_R|^m + K|B_s|$$

i.e.

$$\int_{B_t} |\nabla u|^m dx \le c_2 \left[\int_{B_s \setminus B_t} |\nabla u|^m dx + (s-t)^{-m} \int_{B_R} |u - u_R|^m + K|B_R| \right]. \tag{3.3}$$

Now we fill the hole, i.e. we sum to (3.3) c_2 times the left-hand side and we get

$$\int_{B_t} |\nabla u|^m dx \le \theta \int_{B_s} |\nabla u|^m + c_3 \left[(s-t)^{-m} \int_{B_R} |u - u_R|^m dx + |B_R| \right] \tag{3.4}$$

where

$$\theta = \frac{c_2}{1+c_2} < 1 .$$

Now we would like to eliminate the first term on the right-hand side of (3.4) getting

$$\int_{B_{R/2}} |\nabla u|^m dx \le c_4 \left[R^{-m} \int_{B_R} |u - u_R|^m dx + |B_R| \right] . \tag{3.5}$$

In fact this can be done by means of Lemma 3.1 below; then, through a simple use of the Sobolev-Poincaré inequality, (3.5) gives

$$⨍_{B_{R/2}} (1+|\nabla u|)^m dx \le c_5 \left(⨍_{B_R} (1+|\nabla u|)^q dx \right)^{\frac{m}{q}} \qquad q = \frac{m \cdot n}{m+n} < m$$

and the result follows from Proposition 1.1. q.e.d.

LEMMA 3.1. *Let* $f(t)$ *be a nonnegative bounded function defined in* $[\tau_0, \tau_1]$, $\tau_0 \ge 0$. *Suppose that for* $\tau_0 \le t < s \le \tau_1$ *we have*

$$f(t) \le [A(s-t)^{-\alpha} + B] + \theta f(s) \tag{3.6}$$

where A, B, α, θ *are nonnegative constants with* $0 \le \theta < 1$. *Then for all* $\tau_0 \le \rho < R \le \tau_1$ *we have*

$$f(\rho) \le c\,[A(R-\rho)^{-\alpha} + B] \tag{3.7}$$

where c *is a constant depending on* α *and* θ.

Proof. For ρ and R fixed, let us consider the sequence $\{t_i\}_{i=1,\cdots,\infty}$ defined by

$$t_0 = \rho \qquad t_{i+1} - t_i = (1-\tau)\tau^i (R-\rho)$$

$0 < \tau < 1$.

By iteration from (3.6) we get

$$f(t_0) \le \theta^k f(t_k) + \left[\frac{A}{(1-\tau)^{\alpha}}(R-\rho)^{-\alpha} + B\right] \sum_{0}^{k-1} \theta^i \tau^{-i\alpha} .$$

If we now choose τ in such a way that $\tau^{-\alpha}\theta < 1$ and go to the limit for $k \to \infty$, we get (3.7) with $c = c(\alpha, \theta) = (1-\tau)^{-\alpha}(1-\theta\tau^{\alpha})^{-1}$. q.e.d.

REMARK 3.1. By using Lemma 3.1, one could prove all results in Section 2 using Proposition 1.1 with $\theta = 0$.

REMARK 3.2. Analogous L^q-estimates can be obtained in the same way (compare also the proof of Theorem 2.2), by assuming instead of ii)

$$\mu|p|^m - \gamma|u|^{\sigma} - f^m \le F(x,u,p) \le \Lambda|p|^m + \Gamma|u|^{\sigma} + g^m$$

where

$$\sigma \begin{cases} < \dfrac{mn}{n-m} & \text{if} \quad n > m \\ = \text{any exponent} & \text{if} \quad n = m \end{cases}$$

$$f, g \in L^r(\Omega) \qquad r > m .$$

REMARK 3.3. Going through the proof of Theorem 3.1, it is easy to see that it holds also under the weaker condition that u be a 'strong-quasi-minimum', i.e. there exist two constants ε and c such that

$$J[u] \le c\, J[v]$$

for all $v \in H^{1,m}(\Omega, R^N)$ satisfying $v - u \in H_0^{1,m}(\Omega, R^N)$ and $\|v-u\|_{L^{m^*}(\Omega,R^N)} \le \varepsilon$, m^* being the Sobolev exponent of p. In this case (3.2) holds only for R less than some R_0 depending on u.

As a simple consequence of Sobolev theorem we now have

COROLLARY 3.1. *Suppose* $m = n$. *Then the minimum points of the functional* $J[u]$ *are Hölder-continuous functions.*

This is the well-known result of Morrey [231], Chapter 4.3.

REMARK 3.4. Since it is calculable how much we gain in summability of the gradient (compare with the proof of Proposition 1.1 and Lemma 1.2) we have: *there exists a positive constant* λ_0 *such that if* $m > n - \lambda_0$, *then the minimum points of* $J[u]$ *are Hölder-continuous*, compare also with [300].

REMARK 3.5. As in Section 2, the estimates of this section can be extended up to the boundary, provided Ω is smooth and the boundary value is in suitable L^p-classes.

We conclude this section by presenting the simple proof of Corollary 3.1 due to K.-O. Widman [300] and for the sake of simplicity we assume

$$\mu|p|^2 \leq F(x,u,p) \leq \Lambda|p|^2 + b \,.$$

Let χ be a Lipschitz function, such that $\chi \equiv 0$ on B_R, $\chi \equiv 1$ on ∂B_{2R}, $B_{2R} \subset\subset \Omega$, $0 \leq \chi \leq 1$ on B_{2R}, $|\nabla \chi| \leq c/R$ and let

$$a = \fint_{B_{2R} \setminus B_R} u\, dx \,.$$

For $v = (u-a)\chi + a$, we have

$$\int_{B_{2R}} F(x,u,\nabla u)\, dx \leq \int_{B_{2R}} F(x,v,\nabla v)\, dx$$

i.e.

$$\mu \int_{B_{2R}} |\nabla u|^2 dx \le \int_{B_{2R}\setminus B_R} F(x,v,\nabla\chi(u-\alpha)+\nabla u\cdot\chi) + \int_{B_R} F(x,v,0) \le$$

$$\le \text{const}\left\{ \int_{B_{2R}\setminus B_R} |\nabla\chi|^2|u-\alpha|^2 + \int_{B_{2R}\setminus B_R} |\nabla u|^2 + R^2 \right\}$$

and using Poincaré inequality

$$\int_{B_R} |\nabla u|^2 dx \le K \int_{B_{2R}\setminus B_R} |\nabla u|^2 + cR^2 .$$

Now we may add K times the integral on the left to both sides of the last inequality and divide by $K+1$ to get

$$\int_{B_R} |\nabla u|^2 dx \le \frac{K}{K+1} \int_{B_{2R}} |\nabla u|^2 dx + KR^2$$

which, upon iteration, yields

$$\int_{B_R} |\nabla u|^2 dx \le c\left(\frac{R}{R_1}\right)^{\sigma} \left\{ \int_{B_{2R_1}} |\nabla u|^2 + (2R_1)^{\sigma} \right\}$$

for all R less than some R_1, with $\sigma > 0$ depending on K and k. This inequality obviously implies the result (compare with Chapter III).

Chapter VI

NONLINEAR ELLIPTIC SYSTEMS: THE DIRECT APPROACH TO REGULARITY

In this chapter we want to present the direct approach to the study of partial regularity of solutions of nonlinear systems due to M. Giaquinta, E. Giusti [113][114] and M. Giaquinta, G. Modica [121][122]. It relies ultimately on a perturbation argument like in Chapter III and uses as essential tool the L^p-estimates we have stated in the last chapter. This method allows us to handle quasilinear and nonlinear systems both under controllable and natural quadratic growth conditions.

In general we shall not prove the results in their full generality, as our aim is mainly to show the idea of the proofs. Anyway we shall give a review of them.

1. *Quasilinear systems*: $C^{0,\alpha}$ *and* $C^{1,\alpha}$ *regularity*

Let us begin by considering the simple second order quasilinear system

$$(1.1)\quad \int_\Omega A_{ij}^{\alpha\beta}(x,u)D_\alpha u^i D_\beta \phi^j dx = \int_\Omega [f_i^\alpha(x)D_\alpha\phi^i + f_i\phi^i]dx \qquad \forall\phi \in H_0^1(\Omega,R^N)$$

where we assume

i) $A_{ij}^{\alpha\beta}(x,u)$ *are continuous functions satisfying*

$$(1.2)\quad \begin{cases} |A_{ij}^{\alpha\beta}(x,u)| \le L \\ A_{ij}^{\alpha\beta}(x,u)\xi_\alpha^i\xi_\beta^j \ge \nu|\xi|^2 \qquad \forall\xi;\ \nu>0 \end{cases}$$

ii) $f_i^\alpha \in L^p(\Omega)$ $p > n$; $f_i \in L^q(\Omega)$ $q > n/2$.

Then we have

THEOREM 1.1. *Suppose that i) ii) hold and that* $u \in H^1(\Omega, R^N)$ *be a weak solution to system (1.1). Then there exists an open set* $\Omega_0 \subset \Omega$ *such that* u *is locally Hölder continuous with exponent* $\min\left(1-\frac{n}{p}, 2-\frac{n}{q}\right)$ *in* Ω_0. *Moreover* $\mathcal{H}^{n-s}(\Omega\setminus\Omega_0) = 0$ *for some* $s > 2$.

We have already proved this theorem in Section 1, Chapter IV, assuming for simplicity the coefficients to be uniformly continuous and $f_i^\alpha \equiv f_i \equiv 0$. Here we want to present the direct proof of [113]. In order to illustrate the main idea, we shall distinguish the two cases of continuous and uniformly continuous coefficients.

First, let us assume the coefficients uniformly continuous; then the main ideas are contained in the proof of the following lemma.

LEMMA 1.1. *Under the assumptions of Theorem 1.1, if the coefficients* $A_{ij}^{\alpha\beta}$ *are also uniformly continuous, for every* $x_0 \in \Omega$ *and every* $\rho, R, 0 < \rho < R < \operatorname{dist}(x_0, \partial\Omega) \wedge 1$ *we have the inequality*

$$\int_{B_\rho(x_0)} |\nabla u|^2 dx \le c_1\left[\left(\frac{\rho}{R}\right)^n + \chi(x_0,R)\right] \int_{B_R(x_0)} |\nabla u|^2 dx + c_2 R^{n-2+2\gamma} \tag{1.3}$$

$$\gamma = \min\left(1-\frac{n}{p}, 2-\frac{n}{q}\right)$$

where

$$\chi(x_0, R) = g\left(R + R^{2-n} \int_{B_R(x_0)} |\nabla u|^2 dx\right)$$

$g(t)$ *being a function going to zero as* t *goes to zero, and* c_1, c_2 *are constants.*

Proof. Let $A_{ij0}^{\alpha\beta} = A_{ij}^{\alpha\beta}(x_0, u_{x_0,R})$ and let v be the solution to the Dirichlet problem

$$(1.4)\qquad \begin{cases} -D_\beta(A_{ij0}^{\alpha\beta} D_\alpha v^i) = 0 & j = 1, \cdots, N \quad \text{in} \quad B_{R/2}(x_0) \\ v - u \in H_0^1(B_{R/2}(x_0), R^N) . \end{cases}$$

Then we have, see Section 2, Chapter III, for all $\rho < R/2$

$$\int_{B_\rho(x_0)} |\nabla v|^2 dx \leq c\left(\frac{\rho}{R}\right)^n \int_{B_{R/2}(x_0)} |\nabla v|^2 dx$$

and therefore

$$(1.5)\qquad \int_{B_\rho(x_0)} |\nabla u|^2 dx \leq c\left(\frac{\rho}{R}\right)^n \int_{B_R(x_0)} |\nabla u|^2 dx + c \int_{B_{R/2}(x_0)} |\nabla(u-v)|^2 dx .$$

If we set $w = u-v$, we have $w = 0$ on $\partial B_{R/2}$ and

$$\int_{B_{R/2}(x_0)} A_{ij0}^{\alpha\beta} D_\alpha w^i D_\beta \phi^j dx =$$

$$\int_{B_{R/2}(x_0)} [A_{ij}^{\alpha\beta}(x_0 u_{x_0,R}) - A_{ij}^{\alpha\beta}(x,u)] D_\alpha u^i D_\beta \phi^j + \int_{B_{R/2}(x_0)} (f_i^\alpha D_\alpha \phi^i + f_i \phi^i) dx$$

for every $\phi \in H_0^1(B_{R/2}, R^N)$.

In particular we may take $\phi = w$, so that, using the ellipticity in (1.2) and Hölder inequality, we get

$$\int_{B_{R/2}(x_0)} |\nabla u|^2 dx \le c \int_{B_{R/2}(x_0)} \sum |A^{\alpha\beta}_{ij}(x,u) - A^{\alpha\beta}_{ij0}|^2 |\nabla u|^2 dx +$$

$$(1.6) \qquad + c \int_{B_{R/2}(x_0)} \sum |f^\alpha_i|^2 dx + \left(\int_{B_{R/2}(x_0)} |w|^{2^*} dx \right)^{1/2^*} \cdot$$

$$\cdot \left(\int_{B_{R/2}(x_0)} \sum |f_i|^{\frac{2n}{n+2}} \right)^{\frac{n+2}{2n}} . \quad {}^{1)}$$

Now

$$\int_{B_R} \sum |f^\alpha_i|^2 dx \le \sum \left(\int_{B_R} |f^\alpha_i|^p dx \right)^{2/p} R^{n\left(1-\frac{2}{p}\right)} \le \text{const}\ R^{n-2+2\left(1-\frac{n}{p}\right)}$$

$$\left(\int_{B_R} \sum |f_i|^{\frac{2n}{n+2}} \right)^{\frac{n+2}{n}} \le \sum \left(\int_{B_R} |f_i|^q \right)^{2/q} R^{n+2-\frac{2n}{q}} \le \text{const}\ R^{n-2+2\left(2-\frac{n}{q}\right)} \quad {}^{2)}$$

---

[1] From now on we shall do all the calculations for $n \ge 3$. Simple changes, depending on the Sobolev imbedding theorem, have to be done for $n = 2$.

[2] Note that it would be sufficient to assume instead of ii) that $f^\alpha_i \in L^{2,n-2+2\sigma}(\Omega)$, and $f_i \in L^{\frac{2n}{n+2}, \frac{n}{n+2}(n-2+2\sigma)}$ $\sigma > 0$. (See Chapter III for the definition of $L^{p,\lambda}$.)

and, because of the continuity assumption on $A_{ij}^{\alpha\beta}$, *there exists a non-negative function* $\omega(t)$ *increasing in* t, *concave*[3] *continuous in* $\omega(0) = 0$, *such that for* $x,y \in \Omega$ *and* $u,v \in \mathbf{R}^N$

$$|A_{ij}^{\alpha\beta}(x,u) - A_{ij}^{\alpha\beta}(y,v)| \leq \omega(|x-y|^2 + |u-v|^2) .$$

Therefore we get from (1.6)

$$(1.7) \qquad \int_{B_{R/2}} |\nabla w|^2 dx \leq c \int_{B_{R/2}} \omega^2 |\nabla u|^2 dx + c R^{n-2+2\gamma}$$

$$\omega^2 = \omega^2(R^2 + |u - u_{x_0,R}|^2) .$$

On the other hand using the L^p-estimate (Theorem 2.1, Chapter V) and the boundedness of ω, we have (for some $\sigma > 2$)

$$\int_{B_{R/2}} \omega^2 |\nabla u|^2 dx \leq \left(\int_{B_{R/2}} |\nabla u|^\sigma dx\right)^{2/\sigma} \left(\int_{B_{R/2}} \omega^{\frac{2\sigma}{\sigma-2}} dx\right)^{\frac{\sigma-2}{\sigma}} \leq$$

$$\leq c \int_{B_R} |\nabla u|^2 dx \left(\fint_{B_R} \omega\, dx\right)^{\frac{\sigma-2}{\sigma}}$$

and, as ω is a concave function, we have

$$\fint_{B_R} \omega(R^2 + h)\, dx \leq \omega\left(R^2 + \fint_{B_R} h\, dx\right) .$$

---

[3] i.e. the modulus of continuity can be taken concave. In fact, if $\alpha(t)$ is a continuous and bounded function in $[0, +\infty]$ with $\alpha(0) = 0$, then

$$w(t) = \inf\{\lambda(t) : \lambda \text{ concave and continuous with } \lambda(t) \geq \alpha(t)\}$$

will do.

Therefore we get

$$\int_{B_{R/2}} \omega^2 |\nabla u|^2 dx \le c\,\omega \left(R^2 + \fint_{B_R(x_0)} |u-u_{x_0 R}|^2 dx\right) \int_{B_R} |\nabla u|^2 .$$

Finally, putting together (1.5) and (1.7), with a simple use of Poincaré inequality, we get (1.3) for all $\rho < R/2$. Since (1.3) is obvious for $R/2 < \rho < R$, we get (1.3) for all $\rho < R$. q.e.d.

Proof of Theorem 1.1 (*in case of uniformly continuous coefficients*): Let $x_0 \in \Omega$, $R < \mathrm{dist}(x_0, \partial\Omega) \wedge 1$. Set

$$\phi(x_0, R) = R^{2-n} \int_{B_R(x_0)} |\nabla u|^2 dx .$$

From (1.3) we get for $0 < \tau < 1$

$$\phi(x_0, \tau R) \le c_1[1 + \chi(x_0, R)\tau^{-n}]\tau^2 \phi(x_0, R) + c_2 \tau^{n-2} R^{2\gamma} . \tag{1.8}$$

Let now $\gamma < a < 1$ and choose τ in such a way that

$$2c_1 \tau^{2-2a} = 1$$

since we have

$$\chi(x_0, R) = g(R^2, \phi(x_0, R)) < \tau^n$$

provided R is less than some R_1, and $\phi(x_0, R)$ is less than some ε_1, setting $c_2 \tau^{n-2} = H_0$, we get

$$\phi(x_0, \tau R) \le \tau^{2a} \phi(x_0, R) + H_0 R^{2\gamma} .$$

Therefore by iteration we obtain

$$\phi(x_0, \tau^k R) \le \tau^{2k\alpha}\phi(x_0, R) + H_0(\tau^{k-1}R)^{2\gamma} \sum_{s=0}^{\infty} [\tau^{2(\alpha-\gamma)}]^s \le$$

$$\le \left[\phi(x_0, R) + H_0 \frac{R^{2\gamma}}{\tau^{2\gamma} - \tau^{2\alpha}}\right] \tau^{2k\gamma} < \varepsilon_1$$

provided

$$\phi(x_0, R) < \varepsilon_0 \qquad 2\varepsilon_0 < \varepsilon_1$$

and $R < R_0 < R_1$ is such that

$$H_0 \frac{R^{2\gamma}}{\tau^{2\gamma} - \tau^{2\alpha}} < \varepsilon_0 .$$

We can then conclude: if $\phi(x_0, R) < \varepsilon_0$ for some $R < R_0$, then

$$\phi(x_0, \tau^k R) \le 2\varepsilon_0 \tau^{2k\gamma}$$

and hence for any $\rho < R_0$

$$\phi(x_0, \rho) \le \text{const} \left(\frac{\rho}{R}\right)^{2\gamma} . \tag{1.9}$$

Now the proof goes on as in Theorem 1.1, Chapter IV. Since $\phi(x_0, R)$ is a continuous function of x_0, if $\phi(x_0, R) < \varepsilon_0$ for a point $x_0 \in \Omega$, then there exists a ball $B(x_0, r)$ such that $\phi(x, R) < \varepsilon_0 \; \forall x \in B(x_0, r)$, hence (1.9) holds for every point in $B(x_0, r)$. Then it follows, see Section 1, Chapter III, that u is locally Hölder continuous in $B(x_0, r)$ with exponent γ. In conclusion, there exists an open set $\Omega_0 \subset \Omega$ such that the solution is locally in $C^{0,\gamma}(\Omega_0)$. Moreover Ω_0 is nonvoid and x_0 is regular if and only if

$$\liminf_{R \to 0^+} R^{2-n} \int_{B_R(x_0)} |\nabla u|^2 \, dx = 0$$

i.e.

$$\Omega \setminus \Omega_0 = \left\{ x \in \Omega : \liminf_{R \to 0^+} R^{2-n} \int_{B_R} |\nabla u|^2 dx > 0 \right\}$$

and because of the results in Section 2, Chapter IV, and the L^p-estimate of the last chapter we have

$$\mathcal{H}^{n-s}(\Omega \setminus \Omega_0) = 0$$

for some $s > 2$. q.e.d.

REMARK 1.1. It is worth remarking that $x_0 \in \Omega_0$ if and only if for *some* $R < R_0$ we have

$$R^{2-n} \int_{B_R(x_0)} |\nabla u|^2 dx < \varepsilon_0$$

and that R_0 and ε_0 are explicitly calculable in terms of the data.

Proof of Theorem 1.1 (in case of continuous coefficients): Now let us assume the coefficients to be only continuous instead of uniformly continuous. Then we get exactly as before

$$\int_{B_\rho(x_0)} |\nabla u|^2 dx \leq c\left(\frac{\rho}{R}\right)^n \int_{B_R(x_0)} |\nabla u|^2 + \tag{1.10}$$

$$\int_{B_{R/2}} |A(x,u) - \mathsf{A}(x_0, u_{x_0,R})|^2 |\nabla u|^2 dx + c R^{n-2+2\gamma} .$$

Now there exists a *nonnegative bounded function* $\omega(t,s)$ *increasing in* t *for fixed* s *and in* s *for fixed* t, *concave in* s, *continuous in* $(t,0)$ *with* $\omega(t,0) = 0$, *such that for all* $x,y \in \Omega$ *and for all* p,q *with* $|p| < M$

$$|A(x,p)-A(y,q)| \le \omega(M, |x-y|^2+|p-q|^2) .$$

In fact, this time the modulus of continuity of the A's depends on the point where we freeze the coefficients.[4)]

Then, using as before the boundedness and the concavity of ω and the L^p-estimate, we get

$$\int_{B_{R/2}} |A(x,u)-A(x_0 u_{x_0,R})|^2 |\nabla u|^2 dx \le$$

$$\chi\left(|u_{x_0,R}|, R^2 + R^{2-n} \int_{B_R} |\nabla u|^2 dx\right) \int_{B_R} |\nabla u|^2 dx$$

where $\chi(\tau, t)$ goes to zero for t going to zero uniformly for $|\tau| \le M$, anc taking into account (1.10)

$$\int_{B_\rho(x_0)} |\nabla u|^2 dx \le c\left[\left(\frac{\rho}{R}\right)^n + \chi\left(|u_{x_0,R}|, R^2 + \right.\right.$$

(1.11)

$$\left.\left. + R^{2-n} \int_{B_R(x_0)} |\nabla u|^2 dx\right)\right] \int_{B_R(x_0)} |\nabla u|^2 dx + c R^{n-2+2\gamma} .$$

Now write $\chi(x_0, R)$ instead of

$$\chi\left(|u_{x_0,R}|, R^2 + R^{2-n} \int_{B_R(x_0)} |\nabla u|^2 dx\right) .$$

Since for fixed M, we have

$$\chi(x_0, R) < \tau^n$$

---

4) As we are working locally in Ω we may assume the coefficients uniformly continuous with respect to x in Ω.

provided $|u_{x_0,R}| < M_1$, R is less than some R_1 and $\phi(x_0,R) = R^{2-n}\int_{B_R(x_0)}|\nabla u|^2\,dx$ is less than some ε_1, as before we get

$$\phi(x_0,\tau R) \le \tau^{2\alpha}\phi(x_0,R) + H_1R^{2\gamma}\,.$$

On the other hand

$$|u_{x_0,\tau^k R}| \le |u_{x_0,R}| + H_2(n,N,\tau)\sum_{s=0}^{k}\phi(x_0,\tau^s R)^{1/2}\,.^{5)}$$

Therefore by iteration we get

$$\phi(x_0,\tau^k R) \le \tau^{2k\alpha}\phi(x_0,R) + H_1(\tau^{k-1}R)^{2\gamma}\sum_{s=0}^{\infty}[\tau^{2\alpha-2\gamma}]^s =$$

$$= \left[\phi(x_0,R) + H_1\frac{R^{2\gamma}}{\tau^{2\gamma}-\tau^{2\alpha}}\right]\tau^{2k\gamma} < \varepsilon_1$$

$$|u_{x_0,\tau^k R}| \le |u_{x_0,R}| + H_1\varepsilon_0^{1/2}\sum_{s=0}^{k}\tau^{\gamma s} \le$$

$$\le |u_{x_0,R}| + H_2\varepsilon_0^{1/2}\frac{1}{1-\tau^{\gamma}} \le M_1$$

---

5) Compare with [121][131]. For all $\rho < \operatorname{dist}(x_0,\partial\Omega)$ we have, see the proof of Proposition 1.2, Chapter III:

$$|u_{x_0,\tau\rho} - u_{x_0,\rho}| \le \Big(\int_{B_\rho(x_0)}|u(x)-u_{x_0,\rho}|^2\,dx\Big)^{1/2} +$$

$$+\tau^{-n/2}\Big(\fint_{B_{\tau\rho}(x_0)}|u(x)-u_{x_0,\tau\rho}|^2\,dx\Big)^{1/2} \le$$

$$\le c(n)\Big\{\Big(R^{2-n}\int_{B_\rho(x_0)}|\nabla u|^2\,dx\Big)^{1/2} + \tau^{-n/2}\Big((\tau\rho)^{2-n}\int_{B_{\tau\rho}(x_0)}|\nabla u|^2\,dx\Big)^{1/2}\Big\}.$$

Therefore the estimate follows simply from the inequality

$$|u_{x_0,\tau^k R}| \le |u_{x_0,R}| + \sum_{i=1}^{k}|u_{x_0,\tau^i R} - u_{x_0,\tau^{i-1}R}|\,.$$

provided

$$\phi(x_0, R) < \varepsilon_0 \qquad |u_{x_0,R}| < \frac{M_1}{2} \qquad 2\varepsilon_0 < \varepsilon_1 \qquad R_0 < R_1$$

and provided also that ε_0 and R_0 are chosen in such a way that

$$H_1 \frac{R_0^{2\gamma}}{\tau^{2\gamma} - \tau^{2\alpha}} < \varepsilon_0\,, \qquad H_2 \varepsilon_0^{1/2} \frac{1}{1-\tau^{\gamma}} < \frac{M_1}{2}\,.$$

Now we can conclude: for any M_1, if

$$|u_{x_0,R}| < \frac{M_1}{2} \qquad \phi(x_0, R) < \varepsilon_0(M_1)$$

for some $R < R_0(M_1)$, then

$$\phi(x_0, \tau^k R) \leq 2\varepsilon_0 \tau^{2k\alpha}$$

which implies, as before, the result q.e.d.

REMARK 1.2. Note that $x_0 \in \Omega_0$ if

$$\sup_R |u_{x_0,R}| \leq M < +\infty$$

and if for some $R < R_0(M)$ we have

$$R^{2-n} \int_{B_R(x_0)} |\nabla u|^2 dx < \varepsilon_0(M)$$

where R_0 and ε_0 are explicitly calculable.

Note moreover that this time we have

$$\Omega \backslash \Omega_0 \subset \left\{ x \in \Omega : \liminf_{R \to 0^+} R^{2-n} \int_{B_R(x_0)} |\nabla u|^2 dx > 0 \right\} \cup$$

$$\cup \; \{x \in \Omega : \sup_R |u_{x,R}| = +\infty\}\,.$$

Let us now consider the general quasilinear elliptic system

$$(1.12)\quad \int_\Omega A_{ij}^{\alpha\beta}(x,u)D_\beta u^j D_\alpha \phi^i\,dx = \int_\Omega [a_i^\alpha(x,u)D_\alpha\phi^i + b_i(x,u,\nabla u)\phi^i]\,dx$$

for all $\phi \in C_0^\infty(\Omega, R^N)$, and assume that controllable growth conditions hold. More precisely let us assume that

I_1-(*leading part*): $A_{ij}^{\alpha\beta}(x,u)$ *are continuous functions in* $\Omega\times R^N$, *satisfying*

$$|A_{ij}^{\alpha\beta}(x,u)| \le L$$

$$A_{ij}^{\alpha\beta}(x,u)\xi_\alpha^i\xi_\beta^j \ge \nu|\xi|^2 .$$

I_2-(*lower order terms*): $a_i^\alpha(x,u)$ *and* $b_i(x,u,\nabla u)$ *are measurable for all* $u \in H^1(\Omega, R^N)$ *and the following growth conditions hold*

$$|a_i^\alpha(x,u)| \le \mu_1(|u|^{r/2} + f_{i\alpha})$$

$$|b_i(x,u,\nabla u)| \le \mu_2\Big(|\nabla u|^{2\left(1-\frac{1}{r}\right)} + |u|^{r-1} + f_i\Big)$$

$$r = \begin{cases} \dfrac{2n}{n-2} & \text{if} \quad n>2 \\ \text{any exponent} & \text{if} \quad n=2 \end{cases}$$

$$f_{i\alpha} \in L^p(\Omega) \quad p>n\,;\ f_i \in L^q(\Omega) \quad q>\frac{n}{2} .$$

Then we have, see [121]:

THEOREM 1.2. *Let* I_1 *and* I_2 *hold and let* u *be a weak solution to system (1.12). Then there exists an open set* Ω_0 *such that* $u \in C^{0,\alpha}(\Omega_0)$, $\alpha = \min\left(1-\frac{n}{p}, 2-\frac{n}{q}\right)$; *moreover* $\mathcal{H}^{n-s}(\Omega\setminus\Omega_0) = 0$ *for some* $s>2$.

Proof. Here we only give a sketch of the proof, referring to [121] for more details. We split u as $v+w$ where v is the solution to the Dirichlet problem

$$(1.13)\quad \begin{cases} -D_\beta(A_{ij}^{\alpha\beta}(x_0, u_{x_0,R})D_\alpha v^i) = 0 & j=1,\cdots,N \quad \text{in} \quad B_{R/2}(x_0) \\ v-u \in H_0^1(B_{R/2}(x_0), R^N) . \end{cases}$$

Then we have, as usual,

$$(1.14)\qquad \int_{B_\rho} |\nabla v|^2 dx \le c\left(\frac{\rho}{R}\right)^n \int_{B_{R/2}} |\nabla u|^2$$

and w satisfies

$$\int_{B_{R/2}} A_{ij}^{\alpha\beta}(x_0, u_{x_0,R}) D_\alpha w^i D_\beta \phi^j =$$

$$(1.15)\qquad = \int_{B_{R/2}} [A_{ij}^{\alpha\beta}(x_0, u_{x_0,R}) - A_{ij}^{\alpha\beta}(x,u)] D_\alpha u^i D_\beta \phi^j dx +$$

$$+ \int_{B_{R/2}} [a_i^\alpha(x,u) D_\alpha \phi^i + b_i(x,u,\nabla u)\phi^i] dx$$

for all $\phi \in H_0^1(B_{R/2}, R^N)$. Now we put in (1.15) $\phi = w$, getting as in the proof of Theorem 1.1

$$\int_{B_{R/2}(x_0)} |\nabla w|^2 dx \le \chi\left(|u_{x_0,R}|, R^2 + R^{2-n} \int_{B_R(x_0)} |\nabla u|^2\right) \cdot$$

$$\cdot \int_{B_R} (|\nabla u|^2 + |u|^{\frac{2n}{n-2}}) dx + \int_{B_R} |a_i^\alpha(x,u)|^2 + \text{const}\left(\int_{B_R} |b_i(x,u,\nabla u)|^{\frac{2n}{n+2}} dx\right)^{\frac{n+2}{n}} .$$

Since now

$$\int_{B_R} a_i^\alpha(x,u)|^2 \le \text{const} \int_{B_R} |u|^{\frac{2n}{n-2}} dx + \text{const } R^{n-2+2\gamma}$$

$$\left(\int_{B_R} |b_i(x,u,\nabla u)|^{\frac{2n}{n+2}} dx\right)^{\frac{n+2}{2n}} \le$$

$$\le \text{const}\left(\int_{B_R} (|\nabla u|^2 + |u|^{\frac{2n}{n-2}}) dx\right)^{2/n} \cdot \int_{B_R} (|\nabla u|^2 + |u|^{\frac{2n}{n-2}}) dx + \text{const } R^{n-2+2\gamma}$$

and

$$\int_{B_R} |u|^{\frac{2n}{n-2}} dx \le \text{const} \int_{B_R} |u - u_R|^{\frac{2n}{n-2}} + \text{const } R^n |u_{x_0,R}|^{\frac{2n}{n-2}} \le$$

$$\le \text{const}\left(\int_{B_R} |\nabla u|^2 dx\right)^{\frac{2}{n-2}} \int_{B_R} |\nabla u|^2 + \text{const } R^n |u_{x_0,R}|^{\frac{2n}{n-2}}$$

we finally obtain

$$\int_{B_\rho} (|\nabla u|^2 + |u|^{\frac{2n}{n-2}})\,dx \le \text{const}\left[\left(\frac{\rho}{R}\right)^n + \omega(x_0, R)\right] \cdot$$

$$\cdot \int_{B_R} (|\nabla u|^2 + |u|^{\frac{2n}{n-2}})\,dx + \text{const } R^n |u_{x_0,R}|^{\frac{2n}{n-2}} + \text{const } R^{n-2+2\gamma}$$

where

$$\omega(x_0, R) = \chi\left(|u_{x_0,R}|, R^{2-n} \int_{B_R} |\nabla u|^2\,dx\right) + \left(\int_{B_R} |\nabla u|^2\,dx\right)^{\frac{2}{n-2}} +$$

$$+ \text{const}\left[\int_{B_R} (|u|^{\frac{2n}{n-2}} + |\nabla u|^2)\right]^{2/n} .$$

The result then follows, essentially, as in the proof of Theorem 1.1, see [121]. q.e.d.

As a simple model of the general situation in Theorem 1.2 one can look at the single equation

(1.16) $$-\nabla u = |\nabla u|^\gamma$$

with $\gamma < 1 + \frac{2}{n}$ if $n \ge 3$.

Now we want to consider the case of natural growth conditions, i.e. $1 + \frac{2}{n} < \gamma \le 2$ $(n \ge 3)$, distinguishing for convenience the two cases $1 + \frac{2}{n} < \gamma < 2$ and $\gamma = 2$.

For the sake of simplicity, let us consider weak solutions in Ω of the nonlinear system

(1.17) $$-D_\alpha[A_{ij}^{\alpha\beta}(x,u) D_\beta u^j] = f_i(x, u, \nabla u) \qquad i = s, \cdots, N$$

where

II_1 - (*leading part*). $A^{\alpha\beta}_{ij}(x,u)$ *are continuous functions in* $\Omega \times R^N$ *satisfying*

$$|A^{\alpha\beta}_{ij}(x,u) \leq L(M) \qquad\qquad M = \sup_{\Omega} |u|$$

$$A^{\alpha\beta}_{ij}(x,u)\xi^i_\alpha \xi^j_\beta \geq \nu(M)|\xi|^2 \qquad\qquad \forall\,\xi\,;\ \nu(M) > 0$$

and

II_2 - (*lower order term*). $f(x,u,\nabla u)$ *is measurable for all* $u \epsilon H^1 \cap L^\infty(\Omega, R^N)$ *and*

(a) $|f(x,u,p)| \leq a(M)\,|p|^2 + b$ [6)]

or (*for* $n \geq 3$)

(b) $|f(x,u,p)| \leq a(M)\,|p|^\gamma + b \qquad 1 + \frac{2}{n} < \gamma < 2\,.$

We recall that 'u is a weak solution' means that $u \epsilon H^1 \cap L^\infty(\Omega, R^N)$ and

$$\int_\Omega A^{\alpha\beta}_{ij}(x,u) D_\beta u^j D_\alpha \phi^i dx = \int_\Omega f_i(x,u,\nabla u)\phi^i dx$$

for all $\phi \epsilon H^1_0 \cap L^\infty(\Omega, R^N)$.

We have, see [113][121]

THEOREM 1.3. *Let* u *be a weak solution to system (1.17). Assume that* II_1 *and* II_2*(b) hold, or* II_1, II_2*(a) hold and moreover*

$$(1.18) \qquad\qquad 2a(M)\cdot M < \nu(M)\,.$$

Then there exists an open set $\Omega_0 \subset \Omega$ *such that* $u \epsilon C^{0,\alpha}(\Omega_0, R^N)$ *for every* $\alpha < 1$; *moreover* $\mathcal{H}^{n-q}(\Omega \backslash \Omega_0) = 0$ *for some* $q > 2$.

The proof proceeds exactly as for Theorem 1.1 once we have proved the following

---

6) Instead of $b = \text{const}$ we could assume $b \epsilon L^p\ p > n/2$.

LEMMA 1.2. *With the assumptions of Theorem 1.3, for every* $x_0 \in \Omega$ *and every* ρ, R, $0 < \rho < R < \operatorname{dist}(x_0, \partial\Omega) \wedge 1$, *we have the inequality*

$$\int_{B_\rho(x_0)} (1+|\nabla u|^2)\,dx \le c\left[\left(\frac{\rho}{R}\right)^n + \chi(x_0,R)\right] \cdot \int_{B_R(x_0)} (1+|\nabla u|^2)\,dx \tag{1.19}$$

where

$$\chi(x_0,R) = g\left(R^{2-n}\int_{B_R(x_0)} |\nabla u|^2\,dx\right)$$

$g(t)$ *going to zero as* t *goes to zero.*

Proof. Let v be the solution to the Dirichlet problem

$$\begin{cases} -D_\alpha[A^{\alpha\beta}_{ij}(x_0, u_{x_0,R})D_\beta u^j] = 0 & \text{in } B_{R/2} \\ v = u & \text{on } \partial B_{R/2} \end{cases}$$

then we have

$$\int_{B_\rho(x_0)} |\nabla v|^2\,dx \le \text{const}\left(\frac{\rho}{R}\right)^n \int_{B_R} |\nabla u|^2\,dx \qquad \rho < R/2$$

and (see Proposition 2.3 of Chapter III)

$$\sup_{B_{R/2}} |v| \le \text{const}\, M .$$

Now if we set $w = u - v$, we have $w = 0$ on $\partial B_{R/2}$ and

$$\int A_{ij}^{\alpha\beta}(x_0,u_{x_0,R})D_\beta w^j D_\alpha \phi^i \, dx = \int [A_{ij}^{\alpha\beta}(x_0,u_{x_0,R}) - A_{ij}^{\alpha\beta}(x,u)] \cdot D_\beta u^j D_\alpha \phi^i +$$

$$+ \int f_i \phi^i \qquad \forall \phi \in H_0^1 \cap L^\infty .$$

In particular, we may take $\phi = w$, getting by means of the L^p-estimate, as usual,

$$\int_{B_{R/2}} |\nabla w|^2 dx \le c \int_{B_R} |\nabla u|^2 dx \, \omega\left(R^2 + R^{2-n} \int_{B_R} |\nabla u|^2\right) +$$

$$(1.20) \qquad + \int_{B_{R/2}} (|w|\,|\nabla u|^2 + |w|) dx .$$

As for what concerns the last integral, we have

$$(1.21) \qquad \int_{B_{R/2}} |w|\, dx \le cR \int |\nabla w|\, dx \le cR \left\{ \varepsilon \int_{B_{R/2}} |\nabla w|^2 + \frac{1}{\varepsilon} R^n \right\} .$$

Introducing this inequality in (1.20) and recalling that $R < 1$, ω^2 and w are bounded we get

$$\int_{B_{R/2}} |\nabla w|^2 dx \le c \int_{B_R} (|\nabla u|^2 + 1)\, dx$$

and hence, choosing $\varepsilon = 1$ in (1.21)

$$\int_{B_{R/2}} |w| \le c \cdot R \cdot \int_{B_R} (1 + |\nabla u|^2)\, dx .$$

On the other hand, using again the L^p-estimate of the last chapter

$$\int_{B_{R/2}} |w|\,|\nabla u|^2 \le c \int_{B_R} (1+|\nabla u|^2)\,dx \left(⨍_{B_{R/2}} |w|^2 \right)^{\frac{q-2}{2q}} \le$$

$$\le c \int_{B_R} (1+|\nabla u|^2)\,dx \left(R^2 ⨍_{B_{R/2}} (1+|\nabla u|^2)\,dx \right)^{\frac{q-2}{2q}} .$$

Therefore our result is obtained for $\rho < R/2$, and since it is obvious for $R/2 < \rho < R$, the proof is complete. q.e.d.

REMARK 1.3. It is worth remarking that the smallness condition (1.18), apart maybe from the factor 2, is necessary, as Example 2.1 of Chapter V shows. Moreover, we would like to point out that it enters in the proof of Theorem 1.3 only through the L^p-estimate in Theorem 2.3 of Chapter V.

The conclusions of Theorems 1.2 and 1.3 only under II_1, $II_2(b)$ can be improved to be an *everywhere regularity result*, i.e. $\Omega_0 = \Omega$, if we assume instead of I_1, II_1 that the 'leading part is smooth.' By that we essentially mean that the solutions to system

$$-D_\alpha[A^{\alpha\beta}_{ij} D_\beta u^j] = 0$$

be Hölder continuous. More precisely assume for example that

(i) $A^{\alpha\beta}_{ij} = A^{\alpha\beta}_{ij}(x) \in C^0(\Omega)$ *and the Legendre-Hadamard condition holds* or

(ii) $A^{\alpha\beta}_{ij} = \delta_{ij} A^{\alpha\beta}(x)$, $A^{\alpha\beta}(x) \in L^\infty(\Omega)$ *and*

$$A^{\alpha\beta}\xi_\alpha \xi_\beta \ge \nu|\xi|^2 \qquad \forall \xi;\ \nu > 0$$

or

(iii) $A^{\alpha\beta}_{ij} = A^{\alpha\beta}_{ij}(x) \in L^\infty$, $\nu|\xi|^2 \le A^{\alpha\beta}_{ij}\xi^i_\alpha \xi^j_\beta \le L|\xi|^2$ *and* $\frac{\nu}{L}$ *is near to* 1.

Then we have

THEOREM 1.4. *Assume that (i) or (ii) or (iii) hold, and that I_2 or $II_2(b)$ hold. Then the conclusions of Theorems 1.2, 1.3 hold with* $\Omega_0 = \Omega$.[7)]

Proof. Split $u = v + w$ where v is the solution to the Dirichlet problem

$$\begin{cases} -D_\alpha(A^{\alpha\beta}_{ij0} D_\beta v^j) = 0 & i = 1, \cdots, N \quad \text{in} \quad B_{R/2}(x_0) \\ v - u \in H^1_0(B_{R/2}(x_0), \mathbb{R}^N) & \end{cases}$$

and where

$$A^{\alpha\beta}_{ij0} = \begin{cases} A^{\alpha\beta}_{ij}(x_0) & \text{in situation (i)} \\ L\delta^{\alpha\beta}\delta_{ij} & \text{in situation (ii)}. \end{cases}$$

Suppose now to be under hypothesis $II_2(b)$; then, as in the proof of Lemma 1.2, we get (1.19), where this time $\chi(x_0, R)$ goes to zero or is small enough respectively in the situation (i) or (iii), and the result follows easily.

Let us suppose to be in hypothesis I_2. Then we get

$$\int_{B_\rho(x_0)} (V^2 + |\nabla u|^2)\,dx \le c_1\left[\left(\frac{\rho}{R}\right)^n + \chi(x_0,R)\right] \int_{B_R(x_0)} (V^2 + |\nabla u|^2)\,dx +$$

$$+ c_2 R^{n-2+2\gamma} + c_3 \int_{B_R(x_0)} V^2\,dx$$

$$V^2 = |u|^{\frac{2n}{n-2}}.$$

---

7) For results in the case II(a) we refer to [275].

Now note that $V^2 \in L^{1+\varepsilon}_{\mathrm{loc}}$, $\varepsilon > 0$, then $V \in L^{2,\sigma}_{\mathrm{loc}}$ $(\sigma = \frac{\varepsilon}{1+\varepsilon})$, therefore $|\nabla u| \in L^{2,\sigma}_{\mathrm{loc}}$, and since

$$\int_{B_\rho} |V - V_\rho|^2 dx \leq \mathrm{const}\left(\int_{B_\rho} |\nabla u|^2 dx\right)^{\frac{n}{n-2}}$$

we have

$$V \in L^{2, \frac{n}{n-2}\sigma}.$$

Then step by step we reach the thesis.

The theorem remains to be proved in situation (ii). To do that we split $u = v + w$ as before with

$$A^{\alpha\beta}_{ij0} = \delta_{ij} A^{\alpha\beta}(x)$$

and then it is sufficient to note that from the De Giorgi-Nash theorem one easily gets

$$\int_{B_\rho(x_0)} |\nabla v|^2 dx \leq c\left(\frac{\rho}{R}\right)^{\beta} \int_{B_{R/2}} |\nabla v|^2 dx$$

for some $\beta > n-2$. Then the proof goes on as before (see anyway the proof of Theorem 1.1 in Chapter VII). q.e.d.

REMARK 1.4. For the sake of completeness, it would be worth trying to extend the above regularity results in the case of the natural growth II_2(b) to the weak solutions considered in Remark 2.2 of Chapter V.

Finally, we would like to point out that all the results we have stated hold for higher order systems; we refer to [121] for the statements and the proofs. For part of the result in Theorem 1.4 we refer also to [205].

So far we have proved, in different situations, that the weak solutions to elliptic systems of the type (1.17) are Hölder-continuous everywhere in

Ω or in Ω except for a closed singular set Σ. Now when the coefficients of the system are more than merely continuous, the solution $u(x)$ will show higher regularity in Ω or in $\Omega_0 = \Omega \setminus \Sigma$. To prove that, as we have seen in Chapter II, one usually introduces the solution in the coefficients and in the right-hand side, and then relies on the regularity results for linear systems. However, as the right-hand side of (1.17) shows a dependence on ∇u we first have to prove that $u \in C^{1,\alpha}(\Omega)$ or $C^{1,\alpha}(\Omega_0)$.

This can be done following the method of Ladyzhenskaya Ural'tseva [191]; here we present the simple proof taken from [113], see also [119] [116].

THEOREM 1.5. *Let* $u \in C^{0,\alpha}(\Omega_0)$, *for all* $\alpha < 1$, *be a weak solution to system (1.17). Assume that* II_1 *and* $II_2(a)$ *hold and that the coefficients* $A^{\alpha\beta}_{ij}$ *are Hölder continuous with exponent* σ. *Then the derivatives of* u *are locally Hölder-continuous with the same exponent* σ *in* Ω_0.

Proof. Let $\Omega_1 \subset\subset \Omega_0$, $x_0 \in \Omega_1$ and $R < \frac{1}{2}\operatorname{dist}(x_0, \partial\Omega_0) \wedge 1$. We may split $u = v + w$ in $B(x_0, R)$ as before, and using estimate (2.8) of Chapter III for ∇v we get

$$(1.22)\qquad \int_{B_\rho(x_0)} |\nabla u - (\nabla u)_\rho|^2 dx \le c\left[\left(\frac{\rho}{R}\right)^{n+2} \int_{B_R} |\nabla u - (\nabla u)_R|^2 dx + \int_{B_R} |\nabla w|^2 dx\right].$$

Now, compare with the proof of Lemma 1.2, for some $p > 2$ we have

$$\int_{B_R} |\nabla w|^2 dx \le c\left[R + \left(R^{2-n} \int_{B_R} (1+|\nabla u|^2)\,dx\right)^{\frac{p-2}{2p}} + \right.$$

$$\left. + \,\omega\left(R^2 + cR^{2-n}\int_{B_R} |\nabla u|^2\right)^{\frac{p-2}{p}}\right]\int_{B_R} (1+|\nabla u|^2)\,dx\ .$$

Therefore, since $u \in C^{0,\alpha}(\Omega_0)$ $\forall \alpha < 1$ and, because of the assumptions, we have

$$\omega(t) \le c\, t^{\sigma/2}$$

for the modulus of continuity ω, from (1.22) we get

$$\text{(1.23)}\qquad \begin{aligned} \int_{B_\rho(x_0)} |\nabla u - (\nabla u)_\rho|^2 dx \le c\Bigg\{\left(\frac{\rho}{R}\right)^{n+2} & \int_{B_R(x_0)} [|\nabla u - (\nabla u)_R|^2 dx] + \\ & + R^{n-2+\alpha\left[2+\sigma\frac{p-2}{p}\right]}\Bigg\}\ . \end{aligned}$$

If α is chosen so close to 1 that $\left[2+\sigma\frac{p-2}{p}\right]\alpha > 2$, we may conclude from (1.23) (compare with Chapter III) that ∇u is Hölder-continuous in $\overline{\Omega}$, with some positive exponent. In particular ∇u is bounded. Now, since

$$\int_{B_R} |\nabla w|^2 \le c\left\{\int_{B_R} (\omega^2 + |w|)\,|\nabla u|^2 dx + \int_{B_R} |w|\,dx\right\}$$

we obtain

$$\int_{B_R} |\nabla w|^2 \le c\left\{R^{n+2\sigma} + \int_{B_R} |w|\,dx\right\}\ .$$

The last integral is easily estimated as

$$\int_{B_R} |w| \le (\omega_n R^n)^{1/2} \left(\int_{B_R} |w|^2 \right)^{1/2} \le \text{const } R^{\frac{n+2}{2}} \left(\int_{B_R} |\nabla w|^2 \right)^{1/2}$$

and therefore

$$\int_{B_R} |\nabla w|^2 dx \le \text{const } R^{n+2\sigma} .$$

Introducing the last inequality in (1.22) we get the conclusion. q.e.d.

2. *Nonlinear systems*: $C^{1,\alpha}$ *regularity*

In this section we shall consider general nonlinear elliptic systems of divergence form

$$(2.1) \qquad -\sum_{\alpha=1}^{n} D_\alpha A_i^\alpha(x, u, \nabla u) = B_i(x, u, \nabla u) \qquad i = 1, \cdots, N$$

and we shall suppose

i) $|A_i^\alpha(x, u, \nabla u)| \le L(1 + |\nabla u|)$

ii) $(1 + |p|)^{-1} A_i^\alpha(x, u, p)$ *are Hölder-continuous functions with some exponent* δ *on* $\overline{\Omega} \times R^N$ *uniformly with respect to* p, *i.e.*

$$\limsup_{(\sigma,\tau)\to(0,0)} \frac{A_i^\alpha(x+\sigma, u+\tau, p) - A_i^\alpha(x, u, p)}{(1 + |p|)(|\sigma| + |\tau|)^\gamma} \le k(u) < +\infty$$

iii) $A_i^\alpha(x, u, p)$ *are differentiable functions in* p *with bounded and continuous derivatives*

$$|A^\alpha_{ip^j_\beta}(x, u, p)| \le L$$

iv) *the strong ellipticity condition*

$$A^{\alpha}_{ip^j_\beta}(x, u, p)\xi^i_\alpha \xi^j_\beta \geq \lambda|\xi|^2 \qquad \forall \xi;\ \lambda > 0$$

and

v) *for all* $u \in H^1_{loc} \cap L^\infty(\Omega, R^N)$ $B(x, u, \nabla u)$ *is measurable and* $|B(x, u, \nabla u)| \leq a|p|^2 + b$.

Up to now we have considered only systems with quasilinear leading part. Of course full nonlinearity, as in (2.1), was permitted, but only in the case that it would be possible to reduce the system to a quasilinear one (a fourth order system) via the differentiability theory, compare Chapter II.

This forces us to make an 'unnatural' assumption on the behavior of the derivatives $A^{\alpha}_{iu^j}(x, u, p)$, namely that $A^{\alpha}_{iu^j}(x, u, p)$ has a growth in p of the same order as $A^{\alpha}_{ip^j_\beta}(x, u, p)$, i.e. to suppose that

$$|A^{\alpha}_{iu^j}(x, u, p)| \leq L .$$

We recall that under assumptions i) ... iv), $B \equiv 0$, $H^{1,2}$-solutions are generally not in the space $H^{2,2}_{loc}(\Omega, R^N)$.

In this section we want to present some regularity results for elliptic systems under the 'natural' assumption ii), which corresponds, if A^{α}_i are differentiable in u, to the natural growth condition

$$|A^{\alpha}_{iu^j}(x, u, p)| \leq L(1 + |p|) .$$

These results are due to M. Giaquinta, G. Modica [122], see also P.-A. Ivert [171][172], and they sound as

THEOREM 2.1. *Let* $u \in H^{1,2}(\Omega, R^N)$ *be a weak solution of*

$$\sum_{\alpha=1}^{n} D_\alpha A^{\alpha}_i(x, u, \nabla u) = 0 \qquad i = 1, \cdots, N . \tag{2.2}$$

Suppose that i) ... iv) are satisfied. Then the first derivatives of u *are Hölder-continuous in an open set* Ω_0. *Moreover*

$$\Omega \setminus \Omega_0 \subset \Sigma_1 \cup \Sigma_2 \tag{2.3}$$

where

$$\Sigma_1 = \left\{ x \in \Omega : \liminf_{R \to 0^+} \fint_{B_R(x)} |\nabla u - (\nabla u)_{x,R}|^2 > 0 \right\}$$

$$\Sigma_2 = \{ x \in \Omega : \sup_R (|u_{x,R}| + |(\nabla u)_{x,R}|) = +\infty \} .$$

In particular $\text{meas}\,(\Omega \setminus \Omega_0) = 0$.

THEOREM 2.2. *Let* $u \in H^{1,2} \cap L^\infty(\Omega, R^N)$ *be a weak solution of system (2.1). Suppose that i) ... v) hold and that, if* $|u| \le M$,

$$2aM < \lambda . \tag{2.4}$$

Then the first derivatives of u *are Hölder-continuous in an open set* Ω_0. *Moreover* $\Omega \setminus \Omega_0 \subset \Sigma_1 \cup \Sigma_2$; *in particular* $\text{meas}\,(\Omega \setminus \Omega_0) = 0$.

Actually, higher order and more general nonlinear systems are considered in [122], but here we shall confine ourselves to the simple case of second order systems referring to [122] for the general situation.

The method of the proof is very similar to the one in Section 1, the main new tool being the sharper L^p-estimate for the gradient stated in Lemma 2.1.

Before going into the proof of Theorems 2.1, 2.1 it is worth making a few remarks.

First, as far as the estimate of the dimension of the singular set $\Omega \setminus \Omega_0$ is concerned, we note that from (2.3) and the result in Section 2, Chapter IV it follows that *if* $u \in H^{2,p}(\Omega, R^N)$, $p \ge \frac{2n}{n+2}$, *then* $\mathcal{H}^{n-p+\varepsilon}(\Omega \setminus \Omega_0) = 0$ $\forall \varepsilon > 0$. Let us recall that one has $u \in H^{2,p}$ for some

$p > 2$, if the oscillation of u on small balls is small, for example if u is continuous (or Hölder-continuous) (note that this is true in dimension 2), compare with Chapter II and the end of the next section; of course we have to assume more on the smoothness of A_i^{α} and B_i.

However, we do not know if the estimate of the dimension of the singular set can be improved in general, or if the almost everywhere regularity is optimal.

As for the smallness condition (2.4), in the quasilinear case it is natural apart maybe from the factor 2, as we have seen; but here we show only almost everywhere regularity instead of regularity except on a closed set of zero $(n-2)$-Hausdorff measure; so it is not clear whether it is natural, compare with the results of E. Heinz [152] and M. Grüther [143], who in the very special case of 2-dimensional H-surfaces are able to prove almost everywhere regularity without assuming any smallness condition such as (2.4).

Finally, while Theorem 1.1' permits to answer the problem of the regularity of minimum points for regular multiple integrals of the type

$$\int_{\Omega} F(x,\nabla u)\,dx$$

(see Chapter II) at least when assuming natural growth conditions and in the sense of regularity except on a closed 'small' singular set, we want to remark that Theorems 2.1 and 2.2 leave the regularity problem still completely open for $H^1(\Omega, R^N)$ minimum points of general regular integrals

$$\int_{\Omega} F(x, u, \nabla u)\,dx \,.$$ [8)]

8) See next section and Chapter IX.

Let us now prove Theorem 2.1.[9)]

Set

$$\phi(x_0, R) = \int_{B_R(x_0)} |\nabla u - (\nabla u)_R|^2 \, dx$$

$$A^{\alpha\beta}_{ij0} = A^{\alpha}_{ip^j_\beta}(x_0, u_{x_0,R}, (\nabla u)_{x_0, R/4})$$

$$\tilde{A}^{\alpha\beta}_{ij} = \int_0^1 A^{\alpha}_{ip^j_\beta}(x_0, u_{x_0,R}, (\nabla u)_{x_0,R/4} + t(\nabla u(x) - (\nabla u)_{x_0,R/4}))\, dt \,.$$

Then system (2.2) can be rewritten as

$$(2.5)\quad \begin{aligned} -D_\alpha[A^{\alpha\beta}_{ij0} D_\beta u^j] = & -D_\alpha\{[A^{\alpha\beta}_{ij0} - \tilde{A}^{\alpha\beta}_{ij}][D_\beta u^j - (D_\beta u^j)_{x_0,R/4}]\} - \\ & -D_\alpha\{A^\alpha_i(x_0, u_{x_0,R}, \nabla u(x)) - A^\alpha_i(x, u(x), \nabla u(x))\} \,. \end{aligned}$$

Split u as $v+w$ where v is the solution of the Dirichlet problem

$$\begin{cases} -D_\alpha(A^{\alpha\beta}_{ij0} D_\beta v^j) = 0 & \text{in } B_{R/4}(x_0) \\ v - u \in H^1_0(B_{R/4}(x_0), R^N) \,. \end{cases}$$

Then we have for every $\rho < R/4$, see Chapter III,

$$\int_{B_\rho(x_0)} |\nabla v - (\nabla v)_\rho|^2 \, dx \le c\left(\frac{\rho}{R}\right)^{n+2} \int_{B_{R/4}(x_0)} |\nabla v - (\nabla v)_{R/4}|^2 \, dx$$

---

9) Since the proof of Theorem 2.2 is very similar we shall omit it and refer to [122].

hence

$$\int_{B_\rho(x_0)} |\nabla u-(\nabla u)_\rho|^2 dx \le c\left(\frac{\rho}{R}\right)^{n+2} \int_{B_R(x_0)} |\nabla u-(\nabla u)_R|^2 dx + \tag{2.6}$$
$$+ c \int_{B_{R/4}(x_0)} |\nabla w|^2 dx .$$

Now $w \in H_0^1(B_{R/4}(x_0), R^N)$ satisfies

$$\int_{B_{R/4}} A_{ij0}^{\alpha\beta} D_\beta w^j D_\alpha \phi^i dx = \int_{B_{R/4}} [A_{ij0}^{\alpha\beta} - \tilde{A}_{ij}^{\alpha\beta}] [D_\beta u^j - (D_\beta u^j)_{x_0,R/4}] D_\alpha \phi^i + \tag{2.7}$$
$$+ \int_{B_{R/4}} [A_i^\alpha(x_0,u_R,\nabla u(x)) - A_i^\alpha(x,u(x),\nabla u(x))] D_\alpha \phi^i$$

for any $\forall \phi \in H_0^1(B_{R/4}(x_0), R^N)$. Hence, choosing $\phi = w$, we get

$$\int_{B_{R/4}} |\nabla w|^2 dx \le \int_{B_{R/4}} \Sigma|A_{ij0}^{\alpha\beta} - \tilde{A}_{ij}^{\alpha\beta}|^2 |\nabla u - (\nabla u)_{R/4}|^2 + \tag{2.8}$$
$$+ \int_{B_{R/4}} \Sigma|A_i^\alpha(x_0,u_R,\nabla u(x)) - A_i^\alpha(x,u(x),\nabla u(x))|^2 .$$

Let us estimate the second integral on the right-hand side of (2.8).

From Assumption (ii) it follows that there exist a non-negative bounded and continuous function $\eta(t,s)$ and an increasing function $k(t)$ such that

a) $\eta(t,s)$ is increasing in t for fixed s and in s for fixed t

b) $\eta(t,s)$ is concave in s

c) $\eta(t,0)=0$ and $\eta(t,s) \le k(t)\, s^{\delta/2}$

d) for every $x,y \in \overline{\Omega}$ $u,v \in R^N$, $p \in R^{nN}$ and for every $i=1,\cdots,N$ $\alpha = 1,\cdots,n$ the following inequality holds

$$|A_i^\alpha(x,u,p) - A_i^\alpha(y,v,p)| \le \eta(|u|, |x-y|^2 + |u-v|^2)(1+|p|)\ .$$

For example we can take $\eta(t,s) = k(t)\, s^{\delta/2} \wedge L$. Therefore

$$\int_{B_{R/4}} \Sigma|A_i^\alpha(x_0,u_R,\nabla u(x)) - A_i^\alpha(x,u(x),\nabla u(x))|^2 dx \le$$

$$\le \int_{B_{R/4}} \eta(|u_R|, |x-x_0|^2 + |u(x)-u_R|^2)(1+|\nabla u|)^2 \le$$

$$\le c\left(\int_{B_{R/4}} (1+|\nabla u|)^q dx\right)^{2/q} \left(\int_{B_{R/4}} \eta\right)^{1-2/q} \overset{10)}{\le}$$

$$\le c \int_{B_{R/2}} (1+|\nabla u|^2)\, dx \left(\fint_{B_{R/2}} \eta\, dx\right)^{1-2/q} \le$$

$$\le cR^n \fint_{B_{R/2}} (1+|\nabla u|^2)\, dx\, \eta\left(|u_R|, R^2 + \fint_{B_R} |u-u_R|^2 dx\right)^{1-2/q} \le$$

$$\le k(|u_R|) R^{n+\varepsilon} \left(\fint_{B_R} (1+|\nabla u|^2)\, dx\right)^{1+\varepsilon} \qquad \varepsilon = \delta\left(1-\tfrac{2}{q}\right).$$

---

[10] In fact we have: *let* u *be a weak solution to system (2.2) and let i) iii) and iv) hold. Then there exists a* $q>2$ *such that* $u \in H^{1,q}_{loc}(\Omega, R^N)$ *and for* $x_0 \in \Omega$ *and* $R < \frac{1}{2}\operatorname{dist}(x_0, \partial\Omega)$ *we have*

Noting now that $\int_{B_R} (1+|\nabla u|^2)dx$ can be estimated in terms of $1+|(\nabla u)_R|^2$ and $\phi(x_0, R)$, we obtain

$$\int_{B_{R/4}} \Sigma|A_i^\alpha(x_0,u_R,\nabla u(x)) - A_i^\alpha(x,u,\nabla u(x))|^2 \le \tag{2.9}$$

$$\le H_1(|u_R| + |(\nabla u)_R| + \phi(R)^{1/2}) R^{n+\varepsilon} ,$$

$H_1(t)$ being an increasing function.

Let us estimate the first integral on the right-hand side of (2.8).

From assumption iii) it follows that there exists a nonnegative bounded and continuous function $\omega(t,s)$ such that

a) $\omega(t,s)$ is increasing in t for fixed s and in s for fixed t

b) $\omega(t,s)$ is concave in s for fixed t

c) $\omega(t,0) = 0$

d) for every $(x,u,p)(y,v,q) \in \bar{\Omega}\times R^N \times R^{nN}$ with $|u|+|p| \le M$ and for every $ij = 1,\cdots,N$, $\alpha,\beta = 1,\cdots,n$ the following inequality holds

$$|A^\alpha_{ip^j_\beta}(x,u,p) - A^\alpha_{ip^j_\beta}(y,v,q)| \le \omega(M, |x-y|^2 + |u-v|^2 + |p-q|^2) .$$

Therefore the first integral on the right-hand side of (2.8) is estimated by

$$\int_{B_{R/4}} \omega^2 |\nabla u - (\nabla u)_{R/4}|^2 dx$$

with

$$\omega = \omega(c(n)[1+|u_R| + |(\nabla u)_R| + \phi(x_0,R)^{1/2}], |\nabla u - (\nabla u)_{R/4}|^2) .$$

$$\left(⨍_{B_R(x_0)} (1+|\nabla u|)^q dx\right)^{2/q} \le c ⨍_{B_{2R}(x_0)} (1+|\nabla u|^2) dx$$

compare with Chapter V.

Now we need the following lemma which we shall prove at the end of the section.

LEMMA 2.1. *Let* u *be a weak solution to system (2.2) and let assumption i) ... iv) hold. Then there exist* $\pi > 2$ $(\pi < q)$, $\varepsilon > 0$ *and a constant* c *such that for every* $x_0 \in \Omega$ *and* $R < \operatorname{dist}(x_0, \partial\Omega)$ *the following estimate holds*

$$\left(\fint_{B_{R/4}(x_0)} |\nabla u - (\nabla u)_{R/4}|^{\pi} dx \right)^{2/\pi} \leq c \fint_{B_R(x_0)} |\nabla u - (\nabla u)_R|^2 dx + \tag{2.10}$$
$$+ h(|u_R| + |(\nabla u)_R| + \phi(x_0, R)^{1/2}) R^{\varepsilon}$$

where h(t) *is an increasing function.*

From Lemma 2.1, taking into account the boundedness of ω, we deduce

$$\int_{B_{R/4}} \omega^2 |\nabla u - (\nabla u)_{R/4}|^2 dx \leq \left(\int_{B_{R/4}} |\nabla u - (\nabla u)_{R/4}|^{\pi} dx \right)^{2/\pi} \cdot \tag{2.11}$$
$$\cdot \left(\int_{B_{R/4}} \omega^{\frac{2\pi}{\pi-2}} dx \right)^{1-2/\pi} \leq c \left(\fint_{B_{R/2}} \omega \, dx \right)^{1-2/\pi} \left\{ \int_{B_{R/2}} |\nabla u - (\nabla u)_{R/2}|^2 dx + \right.$$
$$\left. + R^{n+\varepsilon} h(|u_R| + |(\nabla u)_R| + \phi(x_0, R)^{1/2}) \right\} .$$

On the other hand, being ω concave in s, we have

$$\fint_{B_R} \omega \leq \omega(c(n)[1 + |u_R| + |(\nabla u)_R| + \phi(x_0, R)^{1/2}], \phi(x_0, R)) .$$

In conclusion, putting together (2.6), (2.8), (2.9), (2.11) we deduce

LEMMA 2.2. *Let* u *be a weak solution to system (2.2) and let assumptions i)... iv) hold. Then there exist positive constants* c *and* ε *such that for every* $x_0 \in \Omega$ *and* $0 < \rho < R < \min(1, \operatorname{dist}(x_0, \partial\Omega))$ *the following estimate holds*:

$$\int_{B_\rho(x_0)} |\nabla u - (\nabla u)_\rho|^2 dx \leq c\left[\left(\frac{\rho}{R}\right)^{n+2} + \chi(x_0,R)\right] \cdot \int_{B_R(x_0)} |\nabla u - (\nabla u)_R|^2 dx + \tag{2.12}$$

$$+ \Psi(x_0,R) \cdot R^{n+\varepsilon}$$

where

$$\chi(x_0,R) = \omega(|u_R| + |(\nabla u)_R| + \phi(x_0,R)^{1/2}, \phi(x_0,R)^{1/2})$$

$$\Psi(x_0,R) = H(|u_R| + |(\nabla u)_R| + \phi(R)^{1/2})$$

$\omega(t,s)$ *being an increasing function in* t *going to zero uniformly for* $|t| \leq M$ *as* s *goes to zero, and* $H(t)$ *an increasing function.*

Now the proof of Theorem 2.1 follows the lines of the proof of Theorem 1.1.

Proof of Theorem 2.1. Inequality (2.12) can be written, for $0 < \tau < 1$, as

$$\phi(x_0, \tau R) \leq A\tau^2 \phi(x_0,R)[1 + \tau^{-n-2}\chi(x_0,R)] + \Psi(x_0,R)\tau^{-n}R^\varepsilon .$$

Let now $\varepsilon < \sigma < 2$ and τ be chosen in such a way that $2A\tau^{2-\sigma} = 1$. Since for fixed M_1 we have

$$\chi(x_0,R) < \tau^{n+2}$$

provided that $|u_R| + |(\nabla u)_R| \leq M_1$ and $\phi(R)$ be less than some ε_1, setting

$$H_0 = \Psi(M_1)\tau^{-n}$$

we have: if for some R

$$|u_R| + |(\nabla u)_R| < M_1\,, \qquad \phi(R) < \varepsilon_1(M_1)$$

then

$$\phi(x_0, \tau R) \leq \tau^\sigma \phi(x_0,R) + H_0 R^\varepsilon\,.$$

On the other hand for every k

$$|u_{\tau^k R}| + |(\nabla u)_{\tau^k R}| \leq |u_R| + |(\nabla u)_R| + H_1(n,N,\tau) \sum_{s=0}^{k} \phi(x_0, \tau^s R)^{1/2}\,.$$

Therefore by induction we get

$$\phi(x_0, \tau^k R) \leq \tau^{k\sigma} \phi(x_0,R) + H_0 (\tau^{k-1} R)^\varepsilon \sum_{s=0}^{\infty} (\tau^{\sigma-\varepsilon})^s =$$

$$= \phi(x_0,R) + H_0 \frac{R^\varepsilon}{\tau^\varepsilon - \tau^\sigma}\, \tau^{k\varepsilon} < \varepsilon_1$$

$$\varepsilon_1^{1/2} + |u_{\tau^k R}| + |(\nabla u)_{\tau^k R}| \leq |u_R| + |(\nabla u)_R| + H_1 \left(\frac{\varepsilon_1}{2}\right)^{1/2} \cdot \sum_{s=0}^{k} (\tau^{\varepsilon/2})^s +$$

$$+ \varepsilon_1^{1/2} \leq |u_R| + |(\nabla u)_R| + M_1 \left(\frac{\varepsilon_1}{2}\right)^{1/2} \frac{1}{1-\tau^{\varepsilon/2}} + \varepsilon_1^{1/2} \leq M_1$$

if

$$R < R_0,\ \phi(x_0,R) < \frac{\varepsilon_1}{2},\quad |u_R| + |(\nabla u)_R| < \frac{M_1}{2}\,,$$

and if ε_1 and R_0 are chosen in such a way that

$$\varepsilon_1^{1/2} + H_1 \left(\frac{\varepsilon_1}{2}\right)^{1/2} \frac{1}{1-\tau^{\varepsilon/2}} < \frac{M_1}{2}$$

$$H_0 \frac{R^\varepsilon}{\tau^\varepsilon - \tau^\sigma} < \frac{\varepsilon_1}{2} \qquad \text{for} \quad R < R_0\,.$$

Then we can conclude: for any M_1, if

$$(2.13) \qquad |u_R| + |(\nabla u)_R| < \frac{M_1}{2}\,, \quad \phi(R) < \frac{\varepsilon_1}{2}$$

for some $R < R_0(M_1)$, then

$$\phi(x_0, \tau^k R) \le \varepsilon_1 \tau^{k\varepsilon}$$

and hence, for every $\rho < R_0$

$$(2.14) \qquad \phi(x_0, \rho) \le \text{const}\left(\frac{\rho}{R_0}\right)^{\varepsilon}.$$

Now, since $\phi(x_0,R)$ and $|u_R| + |(\nabla u)_R|$ are continuous functions of x_0, if (2.13) holds for a point $x_0 \in \Omega$ then there exists a ball $B(x_0,r)$ such that (2.13) (and therefore (2.14)) holds for every $x \in B(x_0,r)$. Then it follows that the derivatives of u are Hölder-continuous in an open set Ω_0. Obviously one has $\Omega \setminus \Omega_0 \subset \Sigma_1 \cup \Sigma_2$, and the estimate $\text{meas}(\Omega\setminus\Omega_0) = 0$ is a consequence of the Lebesgue Theorem. q.e.d.

Now it remains to prove Lemma 2.1.

Proof of Lemma 2.1. Define

$$G_i^{\alpha}(x,y) = A_i^{\alpha}(y,u(y),\nabla u(x)) - A_i^{\alpha}(x,u(x),\nabla u(x))$$

$$G^2(x,y) = \sum_{i=1}^{N} \sum_{\alpha=1}^{n} |G_i^{\alpha}(x,y)|^2 .$$

Fix $\overline{x} \in \Omega$ and $\rho < \text{dist}(\overline{x}, \partial\Omega)$. For every $\nu = (\nu_{\alpha}^{i})$ and almost every $y \in \Omega$

$$A_i^{\alpha}(y,u(y),\nabla u(x)) - A_i^{\alpha}(y,u(y),\nu) = \sum_{j=1}^{N} \sum_{\beta=1}^{n} \tilde{A}_{ij}^{\alpha\beta}(D_\beta u^j(x) - \nu_\beta^j)$$

where

$$\tilde{A}^{\alpha\beta}_{ij} = \int_0^1 A^{\alpha}_{ip^j_\beta}(y, u(y), \nu + t(\nabla u(x) - \nu))\, dt$$

and, of course,

$$|\tilde{A}^{\alpha\beta}_{ij}| \leq L$$

$$\tilde{A}^{\alpha\beta}_{ij} \xi^i_\alpha \xi^j_\beta \geq \lambda |\xi|^2 \qquad \forall \xi .$$

Hence, for every $\phi \in H^1_0(\Omega, R^N)$ and almost every y we get

$$\int_\Omega \tilde{A}^{\alpha\beta}_{ij}(D_\beta u^j - \nu^j_\beta) D_\alpha \phi^i(x)\, dx = \int_\Omega G^\alpha_i(x,y) D_\alpha \phi^i(x)\, dx . \tag{2.15}$$

Let now $\overline{y} \in \Omega$, and let η be a standard test function on $B\left(\overline{x}, \frac{3}{4}\rho\right)$, $\eta \equiv 1$ on $B\left(\overline{x}, \frac{\rho}{2}\right)$. Inserting $\phi = (u(x) - u_{\overline{x}, \frac{3}{4}\rho} - \nabla u(x) \cdot (x - \overline{x}))\eta^2$ and $\nu = \nabla u(y)$ in (2.15), we deduce the following inequality

$$\int_{B_{\rho/2}(\overline{x})} |\nabla u(x) - \nabla u(y)|^2 dx \leq c\left\{\frac{1}{\rho^2} \int_{B_{\frac{3}{4}\rho}(\overline{x})} |u - u_{\overline{x}, \frac{3}{4}\rho} - \nabla u(y)(x - \overline{x})|^2 dx + \right.$$

$$\left. + \int_{B_\rho(\overline{x})} G^2(x,y) \right\} \leq c\left\{\frac{1}{\rho^2} \int_{B_{\frac{3}{4}\rho}(\overline{x})} |u - u_{\overline{x}, \frac{3}{4}\rho} - (\nabla u)_{\overline{y}, \rho} \cdot (x - \overline{x})|^2 dx + \right.$$

$$\left. + \int_{B_\rho(\overline{x})} G^2(x,y)\, dx + \rho^n |\nabla u(y) - (\nabla u)_{\overline{y}, \rho}|^2 \right\}$$

which, via the Sobolev-Poincaré inequality, by taking the average in y on $B\left(\bar{y},\frac{3}{4}\rho\right)$, becomes

$$\fint_{B_{\frac{\rho}{2}}(\bar{y})} dy \fint_{B_{\frac{\rho}{2}}(\bar{x})} |\nabla u(x)-\nabla u(y)|^2 dx \le c\left\{\left(\fint_{B_{\frac{3}{4}\rho}(\bar{x})} |\nabla u - (\nabla u)_{\bar{y},\frac{3}{4}\rho}|^{\frac{2n}{n+2}} dx\right)^{1+\frac{2}{n}}\right.$$

(2.16)

$$\left. + \fint_{B_{\rho}(\bar{y})} dy \fint_{B_{\rho}(\bar{x})} G^2(x,y)\,dx + \fint_{B_{\frac{3}{4}\rho}(\bar{y})} |\nabla u(y)-(\nabla u)_{\bar{y},\frac{3}{4}\rho}|^2 dy\right\}.$$

On the other hand, inserting $\phi = (u(x)-u_{\bar{x},\rho} - (\nabla u)_{\bar{y},\rho}(x-\bar{x}))\eta^2$ $\eta \in C_0^\infty(B_\rho(\bar{x}))$ $\eta = 1$ on $B\left(\bar{x},\frac{3}{4}\rho\right)$ and $\nu = (\nabla u)_{\bar{y},\rho}$ in (2.15) we deduce

$$\int_{B_{\frac{3}{4}\rho}(\bar{x})} |\nabla u(x)-(\nabla u)_{\bar{y},\rho}|^2 dx \le c\left\{\frac{1}{\rho^2}\int_{B_\rho(\bar{x})} |u-u_{\bar{x},\rho} - (\nabla u)_{\bar{y},\rho}(x-\bar{x})|^2 dx +\right.$$

$$\left. + \int_{B_\rho(\bar{x})} G^2(x,y)\,dx\right\}$$

and by the Sobolev-Poincaré inequality and taking the average in y on $B(\bar{y},\rho)$ we conclude

$$\fint_{B_{\frac{3}{4}\rho}(\bar{x})} |\nabla u - (\nabla u)_{\bar{x},\frac{3}{4}\rho}|^2 dx \le \fint_{B_{\frac{3}{4}}(\bar{x})} |\nabla u(x) - (\nabla u)_{\bar{y},\rho}|^2 dx \le$$

(2.17)

$$\le c\left\{\left(\fint_{B_\rho(\bar{x})} |\nabla u(x) - (\nabla u)_{\bar{y},\rho}|^{\frac{2n}{n+2}}\right)^{1+\frac{2}{n}} + \fint_{B_\rho(\bar{y})} dy \fint_{B_\rho(\bar{x})} G^2(x,y)dx\right\}.$$

Since now for every $p \ge 1$

$$\fint_{B_\rho(\bar{x})} |\nabla u - (\nabla u)_{\bar{y},\rho}|^p dx \le \fint_{B_\rho(\bar{x})} dx \fint_{B_\rho(\bar{y})} |\nabla u(x) - \nabla u(y)|^p dy \tag{2.18}$$

from (2.16) and (2.17) we get

$$\fint_{B_{\frac{\rho}{2}}(\bar{y})} dy \fint_{B_{\frac{\rho}{2}}(\bar{x})} |\nabla u(x) - \nabla u(y)|^2 dx \le$$

(2.19)

$$c\left(\fint_{B_\rho(\bar{y})} dy \fint_{B_\rho(\bar{x})} |\nabla u(x) - \nabla u(y)|^{\frac{2n}{n+2}} dx\right)^{1+\frac{2}{n}} + \fint_{B_\rho(\bar{y})} dy \fint_{B_\rho(\bar{x})} G^2(x,y)dx$$

$$\forall\, \bar{x},\bar{y} \in \Omega, \quad \rho < \operatorname{dist}(\bar{x},\partial\Omega) \cap \operatorname{dist}(\bar{y},\partial\Omega).$$

Therefore, applying Proposition 1.1 of Chapter V in $\Omega\times\Omega$ with

$$g = |\nabla u(x) - \nabla u(y)|^{\frac{2n}{n+2}}, \quad f = |G(x,y)|^{1+\frac{2}{n}}, \quad q = 1+\frac{2}{n}, \quad \theta = 0,$$

we get for some $\pi > 2$, for any $x_0 \in \Omega$ and $R < \operatorname{dist}(x_0,\partial\Omega)$

$$\left(\fint_{B_{R/4}(x_0)} dy \fint_{B_{R/4}(x_0)} |\nabla u(x) - \nabla u(y)|^{\pi} dx \right)^{2/\pi} \le$$

$$\le c \fint_{B_{R/2}(x_0)} dy \fint_{B_{R/2}(x_0)} |\nabla u(x) - \nabla u(y)|^2 +$$

$$+ c\left(\fint_{B_{R/2}(x_0)} dy \fint_{B_{R/2}(x_0)} \cdot G^{\pi}(x,y)\, dx \right)^{2/\pi}$$

which taking into account (2.18) and

$$2 \fint_{B_\rho(\overline{x})} |\nabla u(x) - (\nabla u)_{\overline{x},\rho}|^2 dx = \fint_{B_\rho(\overline{x})} dy \fint_{B_\rho(\overline{x})} |\nabla u(x) - \nabla u(y)|^2 dx$$

becomes

$$\left(\fint_{B_{R/4}(x_0)} |\nabla u - (\nabla u)_{x_0,R}|^{\pi} dx \right)^{2/\pi} \le$$

(2.20)

$$\le c \fint_{B_{R/2}(x_0)} |\nabla u - (\nabla u)_{x_0,R/2}|^2 dx + c\left(\fint_{B_{R/2}(x_0)} dy \fint_{B_{R/2}(x)} G^{\pi}(x,y)\, dx \right)^{2/\pi}$$

In order to get (2.10), it only remains to estimate the last integral in (2.20). Let $\eta(t,s)$ be as before and set

$$\alpha(x) = \eta(|u_R|, |x - x_0|^2 + |u(x) - u_R|^2) .$$

Then we have

$$\left(\fint_{B_{R/2}(x_0)} dy \fint_{B_{R/2}(x_0)} G^{\pi}(x,y)\,dx \right)^{1/\pi} \leq \tag{2.21}$$

$$\leq \left(\fint_{B_{R/2}(x_0)} dy \fint_{B_{R/2}(x_0)} |A_i^{\alpha}(y,u(y),\nabla u(x)) - A_i^{\alpha}(x_0,u_{x_0,R},\nabla u(x))|^{\pi}\,dx \right)^{1/\pi} +$$

$$+ \left(\fint_{B_{R/2}(x_0)} dy \fint_{B_{R/2}(x_0)} |A_i^{\alpha}(x,u(x),\nabla u(x)) - A_i^{\alpha}(x_0,u_{x_0,R},\nabla u(x))|^{\pi}\,dx \right)^{1/\pi} \leq$$

$$\leq \left(\fint_{B_{R/2}(x_0)} a^{\pi}(y)\,dy \fint_{B_{R/2}(x_0)} (1+(\nabla u))^{\pi}\,dx \right)^{1/\pi} +$$

$$+ \left(\fint_{B_{R/2}(x_0)} a^{\pi}(x)(1+|\nabla u(x)|)^{\pi}\,dx \right)^{1/\pi} .$$

Now, from the L^p-estimate and the boundedness of η we deduce

$$\left(\fint_{B_{R/2}} (1+|\nabla u|)^{\pi}\,dx \right)^{1/\pi} \leq c \left(\fint_{B_R} (1+|\nabla u|^2)\,dx \right)^{1/2}$$

$$\left(\fint_{B_{R/2}} a^{\pi}(y)\,dy \right)^{1/\pi} \leq c \left(\fint_{B_{R/2}} a(y)\,dy \right)^{1/\pi}$$

while, since $\eta(t,s)$ is concave in s and $\eta(t,0) = 0$, $\eta(t,s) \leq k(t)s^{\delta/2}$, using Poincaré inequality we get

$$\fint_{B_{R/2}} \alpha(y)\,dy \le \eta\left(|u_R|, R^2 + \fint_{B_R} |u-u_R|^2\,dx\right) \le$$

$$\le k(|u_R|)R^{\delta}\left(\fint_{B_R}(1+|\nabla u|^2)\,dx\right)^{\delta} .$$

On the other hand, if $\pi < s < q$, q being the exponent in the L^p-estimate for $|\nabla u|$, we have

$$\fint_{B_{R/2}} \alpha^{\pi}(x)(1+|\nabla u(x)|)^{\pi}\,dx \le c\left(\fint_{B_{R/2}}(1+|\nabla u|)^s\,dx\right)^{\pi/s}\left(\fint_{B_{R/2}}\alpha\,dx\right)^{1-\pi/s} \le$$

$$\le c\left(\fint_{B_R}(1+|\nabla u|^2)\,dx\right)^{\pi/2}\left(\fint_{B_R}\alpha\,dx\right)^{1-\pi/s} .$$

Hence, the result follows from (2.20) and (2.21) because of the estimates of the terms on the right-hand side of (2.21) and noting that $\fint_{B_R}(1+|\nabla u|^2)dx$ can be estimated in terms of $1+|(\nabla u)_R|^2$ and $\phi(x_0,R)$. q.e.d.

3. *Minima of quadratic multiple integrals*: $N \ge 1$

As we have remarked, the results in Section 2, although quite general, do not cover the case of minima of regular multiple integrals in the calculus of variations.

Let us consider a regular multiple integral

$$J[u;\Omega] = \int_{\Omega} F(x,u,\nabla u)\,dx \tag{3.1}$$

with

$$(3.1)' \qquad |p|^2 - k \le F(x,u,p) \le a|p|^2 + k$$

and let $u \in H^{1,2}_{loc}(\Omega, R^N)$ be a *local minimum point,* i.e. let us assume that for all $\phi \in H^{1,2}(\Omega, R^N)$ with $\operatorname{spt}\phi \subset\subset \Omega$ we have

$$J[u;\operatorname{spt}\phi] \le J[u+\phi;\operatorname{spt}\phi]\,.$$

Then no previous result answers the question of the partial regularity of u.[11)]

Here we want to present a contribution[12)] to this problem due to M. Giaquinta, E. Guisti [114]. It refers to the special case of quadratic functionals, i.e. multiple integrals of the type

$$(3.2) \qquad \int_\Omega A^{\alpha\beta}_{ij}(x,u)\, D_\alpha u^i D_\beta u^j\, dx \qquad (A^{\alpha\beta}_{ij} = A^{\beta\alpha}_{ji})\,.$$

Let us assume that the coefficients $A^{\alpha\beta}_{ij}$ are bounded continuous functions in $\Omega \times R^N$ and satisfy the Legendre-Hadamard condition

$$(3.3) \qquad A^{\alpha\beta}_{ij}(x,u)\,\xi_i\xi_j\eta_\alpha\eta_\beta \ge \lambda|\xi|^2|\eta|^2 \qquad \forall \xi,\eta;\ \lambda > 0\,.$$

Moreover we suppose that the conclusion of Theorem 3.1, Chapter V holds for the functional (3.2). Of course this is true if inequalities (3.1)′ hold, but this does not seem to follow from (3.3). Inequalities (3.1)′ would instead follow from

$$A^{\alpha\beta}_{ij}(x,u)\,\xi^i_\alpha\xi^j_\beta \ge \lambda|\xi|^2 \qquad \forall \xi;\ \lambda > 0\,.$$

---

11)As we know u is generally not everywhere regular, compare with Section 3 Chapter II. Let us again note that the functional J is not differentiable in $H^{1,2}$, and that we are not allowed to think of u as a solution of the Euler equation, compare with Section 5 of Chapter I.

12)For more results we refer to Chapter IX.

We shall prove the following theorem, where for the sake of simplicity, we shall assume that the coefficients $A_{ij}^{\alpha\beta}$ be uniformly continuous and bounded in $\Omega \times R^N$.[13)]

THEOREM 3.1. *Let the hypothesis above be satisfied, and let* $u \in H^{1,2}_{loc}(\Omega, R^N)$ *be a local minimum point for the functional in (3.2). Then there exists an open set* $\Omega_0 \subset \Omega$ *such that* $u \in C^{0,\alpha}(\Omega_0, R^N)$ *for every* $\alpha < 1$. *Moreover we have*

$$\Omega \backslash \Omega_0 = \left\{ x_0 \in \Omega : \liminf_{R \to 0^+} R^{2-n} \int_{B_R(x_0)} |\nabla u|^2 dx > \varepsilon_0 \right\} \tag{3.4}$$

where ε_0 *is a positive constant independent of* u. *Finally*

$$\mathcal{H}^{n-q}(\Omega \backslash \Omega_0) = 0$$

for some $q > 2$.

Proof. Let $x_0 \in \Omega$, $R < \frac{1}{2} \operatorname{dist}(x_0, \partial\Omega)$, and let v be the solution of the variational problem

$$\begin{cases} \displaystyle\int_{B_R(x_0)} A_{ij}^{\alpha\beta}(x_0, u_R) D_\alpha v^i D_\beta v^j \, dx \to \min \\ v - u \in H_0^1(B_R(x_0), R^N) . \end{cases}$$

Since the coefficients are now constant, the Euler operator is coercive and the problem has a unique solution. Moreover, we have

---

13)This implies, as we have seen, that there exists a continuous, increasing, function $\omega : R^+ \to R^+$ satisfying $\omega(0) = 0$, $\omega(t) \leq M$, and such that

$$|A_{ij}^{\alpha\beta}(x, u) - A_{ij}^{\alpha\beta}(y, v)| \leq \omega(|x-y|^2 + |u-v|^2)$$

and moreover ω is concave.

$$(3.5)\qquad \int_{B_R(x_0)} |\nabla v|^p \, dx \le c_1 \int_{B_R(x_0)} |\nabla u|^p \, dx \quad ^{14)}$$

and for every $\rho < R$, compare Chapter III

$$(3.6)\qquad \int_{B_\rho(x_0)} |\nabla v|^2 \, dx \le c_2 \left(\frac{\rho}{R}\right)^n \int_{B_R(x_0)} |\nabla v|^2 \, dx \, .$$

Set now $w = u - v$; we have $w \in H_0^{1,2}(B_R, R^N)$ and therefore

$$c_3 \int_{B_R} |\nabla w|^2 \, dx \le \int_{B_R} A_{ij}^{\alpha\beta}(x_0, u_R) D_\alpha w^i D_\beta w^j \, dx \, .$$

On the other hand

$$\int_{B_R} A_{ij}^{\alpha\beta}(x_0, u_R) D_\alpha v^i D_\beta w^j \, dx = 0$$

and therefore

$$\int_{B_R} A_{ij}^{\alpha\beta}(x_0, u_R) D_\alpha w^i D_\beta w^j \, dx = \int_{B_R} A_{ij}^{\alpha\beta}(x_0, u_R) D_\alpha u^i D_\beta w^j \, dx =$$

$$= \int_{B_R} [A_{ij}^{\alpha\beta}(x_0, u_R) - A_{ij}^{\alpha\beta}(x, u)] D_\alpha (u^i + v^i) D_\beta w^j dx + \int_{B_R} [A_{ij}^{\alpha\beta}(x, v) - A_{ij}^{\alpha\beta}(x, u)] D_\alpha v^i D_\beta v^j dx +$$

$$+ \int_{B_R} A_{ij}^{\alpha\beta}(x, u) D_\alpha u^i D_\beta u^j - \int_{B_R} A_{ij}^{\alpha\beta}(x, v) D_\alpha v^i D_\beta v^j \, dx \, .$$

---

14) This follows either from the L^p-theory for elliptic operators with constant coefficients, or from the results in Chapter V.

Since u minimizes J, and u = v on ∂B_R, the sum of the last two terms is nonpositive. Therefore

$$\int_{B_R} |\nabla w|^2 dx \le c_4 \int_{B_R} [|\nabla u|^2 + |\nabla v|^2][\omega^2(R^2 + |u-u_R|^2) + \omega^2(R^2 + |u-v|^2)] dx .$$

Taking into account the boundedness of ω and the L^p-estimate in Theorem 3.1, Chapter V, we deduce

$$\int_{B_R} |\nabla u|^2 \omega^2 dx \le c_3 \left(\int_{B_R} |\nabla u|^q dx\right)^{2/q} \left(\int_{B_R} \omega\, dx\right)^{1-2/q} \le$$

$$\le c_6 \int_{B_{2R}} (1 + |\nabla u|^2) dx \left(\fint_{B_R} \omega\, dx\right)^{1-2/q}$$

and using (3.5) with $p = q$

$$\int_{B_R} |\nabla v|^2 \omega^2 dx \le c_6 \int_{B_{2R}} (1 + |\nabla u|^2) dx \left(\fint_{B_R} \omega\, dx\right)^{1-2/q} .$$

Since ω is concave, we have

$$\fint_{B_R} \omega(R^2 + |u-v|^2) dx \le \omega\left(R^2 + \fint_{B_R} |u-v|^2 dx\right) \le$$

$$\le \omega\left(R^2 + c_7 R^{2-n} \int_{B_R} |\nabla w|^2 dx\right) \le \omega\left(R^2 + c_8 R^{2-n} \int_{B_R} |\nabla u|^2 dx\right)$$

and similarly

$$\fint_{B_R} \omega(R^2 + |u-u_R|^2)\,dx \le \omega\, R^2 + c_8 R^{2-n} \int_{B_R} |\nabla u|^2\,dx \ .$$

In conclusion

$$\int_{B_R} |\nabla w|^2\,dx \le c_9 \omega \Big(R^2 + c_{10} R^{2-n} \int_{B_R} |\nabla u|^2\,dx\Big)^{1-2/q} \int_{B_{2R}} (1 + |\nabla u|^2)\,dx$$

and from (3.6)

$$\int_{B_\rho} (1 + |\nabla u|^2)\,dx \le$$

$$\le c_{11}\Big[\Big(\frac{\rho}{R}\Big)^n + \omega\Big(R^2 + c_{10} R^{2-n} \int_{B_R} |\nabla u|^2\,dx\Big)^{1-\frac{2}{q}}\Big] \cdot \int_{B_{2R}} (1 + |\nabla u|^2)\,dx$$

for every $\rho < R < \frac{1}{2}\,\mathrm{dist}\,(x_0, \partial\Omega)$.

The result now follows as in the proof of Theorem 1.1. q.e.d.

REMARK 3.1. By adapting the proof given above, compare also with Theorem 1.5, it is not difficult to show that if the coefficients $A_{ij}^{\alpha\beta}$ are Hölder-continuous, then the first derivatives of the minimum point u are locally Hölder-continuous in Ω_0.

REMARK 3.2. A similar theorem has been proved by R. Schoen-K. Uhlembeck [261] for harmonic maps between manifolds minimizing the energy functional.

The case of (nonuniformly) continuous coefficients, as we know, needs some technical adjustments both in the statement and in the proof. We shall leave the details to the reader.

PROBLEM. Does the result of Theorem 3.1 hold for minima of general regular multiple integrals (of the type (3.1)) in the calculus of variations?

Closely related to the above problem is the following question.

PROBLEM. Let us consider a weak solution u to the elliptic system

$$-D_\alpha A_i^\alpha(\nabla u) = 0 . \tag{3.7}$$

Is it true that there exists a number $\varepsilon_0 > 0$ such that whenever $R^{2-n} \int_{B_R(x_0)} |\nabla u|^2 dx < \varepsilon_0$ then x_0 is a regular point for u, i.e. u is Hölder-continuous in a neighborhood of x_0 ?[15)]

We conclude this section and this chapter with some remarks on the two-dimensional case.

From now on assume $n = 2$. Under the assumptions of Theorem 2.1 or 2.2 we have $u \in H^{1,p}_{loc}$ for some $p > 2$, then, by Sobolev theorem, u is Hölder-continuous. Therefore, see Theorem 1.2, Chapter II, we have $u \in H^{2,2}_{loc}(\Omega, R^N)$. More precisely, choosing in the proof of Theorem 1.2, Chapter II, first $\phi = D_s[(D_s u - \xi)\eta^4]$ and then $\phi = (u - u_{2R})|\nabla u|^2\eta^4$ we deduce

$$\int |\nabla u|^4 \eta^4 dx + \int |\nabla^2 u|^2 \eta^4 dx \leq \int [\eta^4 + |u - u_{2R}|^4 |\nabla \eta|^4 + |\nabla u - \xi|^2 |\nabla \eta|^2 \eta^2] dx . \tag{3.8}$$

Now, taking $\xi = (\nabla u)_{x_0, R}$, we have, using Sobolev-Poincaré inequality

---

15)Note that the equation in variation of (3.7) is a quasilinear system in the derivatives; therefore, from Theorem 1.1, we may deduce that the derivatives of u are Hölder-continuous in a neighborhood of x_0 whenever

$$\fint_{B_R(x_0)} |\nabla u - (\nabla u)_{x_0, R}|^2 dx$$

is small enough.

$$\fint |\nabla\eta|^2|\nabla u-\xi|^2 \leq c\Big(\fint_{B_{2R}} |\nabla^2 u|^{\frac{2n}{n+2}}dx\Big)^{1+\frac{2}{n}}$$

$$\fint |\nabla\eta|^4|u-u_{2R}|^2 \leq c\Big(\fint_{B_{2R}} |\nabla u|^{\frac{4n}{n+4}}dx\Big)^{1+\frac{4}{n}} \leq c\Big(\fint_{B_{2R}} |\nabla u|^{\frac{4n}{n+2}}dx\Big)^{1+\frac{2}{n}}$$

and from (3.8)

$$\fint_{B_R(x_0)} [1+|\nabla u|^{\frac{4n}{n+2}}+|\nabla^2 u|^{\frac{2n}{n+2}}]^{\frac{n+2}{n}}dx \leq$$

$$\leq c\Big(\fint_{B_{2R}(x_0)} [1+|\nabla u|^{\frac{4n}{n+2}}+|\nabla^2 u|^{\frac{2n}{n+2}}]dx\Big)^{\frac{n+2}{n}} .$$

This, through Proposition 1.1, Chapter V, gives that $u \in H^{2,p}_{loc}$ for some $p > 2$. Hence we can conclude

PROPOSITION 3.1. *Under the assumptions of Theorem 2.1 or 2.2 of this chapter, if* $n = 2$, *then the derivatives of weak solutions are Hölder-continuous everywhere.*

Since the local minimum points $u \in H^{1,2}$ of the regular multiple integrals (3.1), (3.1)′ are Hölder-continuous in dimension 2, see Section 3 of Chapter V, then we obtain (compare with the proof of Theorem 1.5).

PROPOSITION 3.2. *The local minimum points of the functional (3.1), (3.1)′ are, if* $n = 2$, C^1*-Hölder-continuous. Therefore, if* F *is analytic, the local minimum points are analytic functions* (*if* $n = 2$).

Chapter VII

NONLINEAR ELLIPTIC SYSTEMS: SPECIAL STRUCTURES AND EVERYWHERE REGULARITY

In Chapter IV we stated the general problem of finding reasonable conditions under which weak solutions of nonlinear elliptic systems are everywhere regular.

In this chapter we shall consider a few situations in which it is possible to show regularity everywhere.

1. *Single equations*

Let us consider a nonlinear elliptic second order equation in divergence form

$$-D_\alpha a_\alpha(x,u,\nabla u) + b(x,u,\nabla u) = 0 \tag{1.1}$$

under natural growths, i.e. we assume: the functions $a_\alpha(x,u,p)$ and $b(x,u,p)$ are measurable in x and continuous in (u,p); $a_\alpha(x,u,p)$ are differentiable in p; moreover

$$\begin{cases} |a_\alpha(x,u,p)| \leq c(1+|p|) \\ |b(x,u,p) \leq c(1+|p|^2) \\ \left|\dfrac{\partial a_\alpha}{\partial p_\beta}(x,u,p)\right| \leq L \end{cases} \tag{1.2}$$ [1)]

and the ellipticity condition holds:

[1)] These assumptions could be slightly weakened, but that would introduce only technical difficulties.

$$(1.3) \qquad \frac{\partial a_\alpha}{\partial p_\beta} \xi_\alpha \xi_\beta \geq \nu |\xi|^2 \qquad \forall \xi ; \; \nu > 0 .$$

Here L and c may depend on $M = \sup |u|$.

A well-known result due to O. A. Ladyzhenskaya and N. N. Ural'tseva [190] states:[2)]

THEOREM 1.1. *Under the assumption (1.2), (1.3), all bounded solutions to equation (1.1) are Hölder-continuous.*[3)]

Here we want to present a different proof, which relies on the partial regularity. The proof's method is taken from [109] [114], see also [296].

Proof of Theorem 1.1. In order to underline the main points, let us divide the proof in three steps:

I. Under our assumptions, see Proposition 2.1 of Chapter V,[4)] we have $u \in H^{1,q}_{loc}(\Omega)$ for some $q > 2$ and

$$\left(\fint_{B_{R/2}} (1 + |\nabla u|)^q \, dx \right)^{1/q} \leq c \left(\fint_{B_R} (1 + |\nabla u|^2) dx \right)^{1/2}$$

for $B_R \subset \Omega$.

[2)] O. A. Ladyzhenskaya and N. N. Ural'tseva use an extension of De Giorgi's technique [69]; a different proof is due to N. S. Trudinger (compare [129]) who uses Moser's technique [234] [235].

For results under controllable growth conditions we refer to [267], see also [283] and [227] [231].

[3)] We recall that the boundedness condition is necessary, compare Section 3 of Chapter II.

[4)] We note that from (1.3), (1.2) it follows that

$$a_\alpha(x,u,p) \, p_\alpha \geq \nu |p|^2 - \text{const}$$

which is the hypothesis actually needed.

II. By the simple standard device of setting

$$a_\alpha(x,u,p) - a_\alpha(x,u,0) = \int_0^1 a_{\alpha p_\beta}(x,u,tp)\,dt \cdot p_\beta$$

we can reduce to the case where in (1.1)

$$a_\alpha(x,u,\nabla u) = A_{\alpha\beta}(x) D_\beta u$$

with

$$A_{\alpha\beta}(x) \in L^\infty(\Omega) \qquad A_{\alpha\beta}\xi_\alpha\xi_\beta \geq \nu|\xi|^2 \qquad \forall \xi;\ \nu > 0\,.$$

Now split u as $u = v + (u-v)$ where v is the weak solution to the Dirichlet problem

$$\begin{cases} \displaystyle\int_{B_R(x_0)} A_{\alpha\beta}(x) D_\beta v D_\alpha \phi = 0 & \forall \phi \in H_0^1(B_R(x_0)) \\ v - u \in H_0^1(B_R(x_0))\,. \end{cases}$$

Then we have from De Giorgi-Nash's theorem

$$\int_{B_\rho(x_0)} |\nabla v|^2 dx \leq c\left(\frac{\rho}{R}\right)^{n-2+2\gamma} \int_{B_R(x_0)} |\nabla v|^2 dx$$

for all $\rho < R$ and for some $\gamma > 0$. Moreover $u - v$, which is bounded since u and therefore v are bounded, satisfies

$$\int_{B_R} A_{\alpha\beta}(x) D_\beta(u-v) D_\alpha \phi = \int b(x,u,\nabla u)\,\phi \qquad \forall \phi \in H_0^1(B_R) \cap L^\infty(B_R)$$

hence, using step I and the boundedness of u, we get

$$\int_{B_R} |\nabla(u-v)|^2 dx \le c \int (|u-v|\,|\nabla u|^2 + |u-v|)dx \le$$

$$\le c \int_{B_{2R}} (1+|\nabla u|^2)\,dx \left(R^2 \fint_{B_{2R}} (1+|\nabla u|^2)\,dx\right)^{\frac{q-2}{2q}} .$$

Now, as in the proof of Theorem 1.1, Chapter VI, it follows that u is Holder-continuous in an open set Ω_0 and

$$\Omega \setminus \Omega_0 = \left\{ x \in \Omega : \liminf_{R \to 0} R^{2-n} \int_{B_R(x_0)} |\nabla u|^2 dx > \varepsilon_0 \right\} .$$

III. The final step is now to show that $\Omega_0 = \Omega$ i.e. that for every x_0 there exists ρ such that

$$\rho^{2-n} \int_{B_\rho(x_0)} |\nabla u|^2 dx \le \varepsilon_0 . \tag{1.4}$$

Choosing as test function $\phi = ue^{t|u|^2}\eta$ we easily get

$$\int_{B_{2R}(x_0)} A_{\alpha\beta} D_\beta \frac{|u|^2}{2} D_\alpha \eta e^{t|u|^2} dx + t \int_{B_{2R}} A_{\alpha\beta} D_\alpha u \cdot u D_\alpha |u|^2 \eta e^{t|u|^2} dx +$$

$$+ \int_{B_{2R}} A_{\alpha\beta} D_\beta u D_\alpha u e^{t|u|^2} \eta \le c \left\{ \int_{B_{2R}} |\nabla u|^2 |u| e^{t|u|^2} + \int_{B_{2R}} \eta \right\} \le$$

$$\le C_1 \left\{ \int_{B_{2R}} |\nabla u|\, |\nabla |u|^2| e^{t|u|^2} \eta + \int_{B_{2R}} \eta \right\}$$

and choosing t large enough

$$\int_{B_{2R}(x_0)} |\nabla u|^2 e^{t|u|^2} \eta\, dx \le -c_2 \int_{B_{2R}} A_{\alpha\beta} e^{t|u|^2} D_\beta |u|^2 D_\alpha \eta + \tag{1.5}$$

$$+ c_3 \int_{B_{2R}} \eta\, dx \qquad c_2, c_3 > 0 .$$

Therefore the function $z = M(2R) - |u|^2$, $M(t) = \sup_{B_t(x_0)} |u|^2$, is a non-negative supersolution for an elliptic operator with right-hand side

$$\int_{B_{2R}} A_{\alpha\beta} e^{t|u|^2} D_\beta z D_\alpha \eta\, dx \ge \frac{c_3}{c_2} \int_{B_{2R}} \eta \qquad \forall \eta \in C_0^\infty(B_{2R}) \quad \eta \ge 0 ,$$

and from the weak Harnack inequality [235] [129] we have

$$R^{-n} \int_{B_{2R}(x_0)} z\, dx \le c_4 [\inf_{B_R} z + R^2] . \tag{1.6}$$

Now let $w \in H_0^1(B_{2R}(x_0))$ be the solution of the equation

$$\int_{B_{2R}} A_{\alpha\beta} e^{t|u|^2} D_\alpha w D_\beta \phi = \frac{1}{R^2} \int_{B_{2R}} \phi\, dx \qquad \forall \phi \in H_0^1(B_{2R}) .$$

Taking $\phi = wz$ we get

$$\frac{1}{2} \int_{B_{2R}} A_{\alpha\beta} e^{t|u|^2} D_\alpha w^2 D_\beta z + \int_{B_{2R}} z A_{\alpha\beta} e^{t|u|^2} D_\alpha w D_\beta w = \frac{1}{R^2} \int_{B_{2R}} wz .$$

The second integral on the left-hand side is nonnegative; moreover we have $w \le \alpha$ in B_{2R} and, from (1.6), $w \ge \alpha_2 > 0$ in B_R, since w is a positive supersolution (α_1 and α_2 do not depend on R). In conclusion, taking $\eta = w^2$, we get

$$\int_{B_{2R}(x_0)} A_{\alpha\beta} e^{t|u|^2} D_\beta z D_\alpha \eta \, dx \le c_5 R^{-2} \int_{B_{2R}} z \, dy$$

which together with (1.5) and (1.6) gives

$$(1.7) \qquad \int_{B_R} |\nabla u|^2 dx \le c_6 R^{n-2}[\inf_{B_R} z + R^2] = c_6 R^{n-2}[(M(2R) - M(R)) + R^2] .$$

On the other hand, we have

$$(1.8) \qquad \sum_{k=0}^{\infty} [M(2^{1-k}R) - M(2^{-k}R)] \le M(2R) \le \sup_{\Omega} |u|^2$$

and inequality (1.7) implies immediately (1.4) with $\rho = 2^{-k}R$ for some k and therefore the regularity of u . q.e.d.

REMARK 1.1. We note that from (1.8) it follows that the radius ρ for which (1.4) holds can be estimated only in terms of $\sup_{\Omega} |u|$ and hence the Hölder norm of u in any relatively compact set $K \subset \Omega$ is bounded in terms of dist $(K, \partial\Omega)$ and $\sup |u|$.

Assume now that $a_\alpha(x,u,p)$ *are Hölder-continuous functions with exponent* δ *in* (x,u) *uniformly with respect to* p *and differentiable in* p; $b(x,u,p)$ *is measurable in* x *and continuous in* (u,p) *and finally (1.2) and (1.3) still hold.*

Then we have

THEOREM 1.2. *The bounded weak solutions to (1.1) belong to* $C^{1,\delta}$.

Proof. The proof we give is a readjustment of the proof of Theorem 1.5 in Chapter VI; for a different one see [191]. First let us make two simple remarks. Assume that $a_\alpha = a_\alpha(p)$ with a_α differentiable in p (growth and ellipticity conditions remaining unchanged), and consider a weak solution v to

$$\int_\Omega a_\alpha(\nabla v)\, D_\alpha \phi\, dx = 0 \qquad \forall \phi \in H_0^1(\Omega)\,.$$

Differentiating we get $D_s v \in H^1_{loc}(\Omega)$ for $s = 1, 2, \cdots, n$ and

$$\int a_{\alpha p_\beta} D_\beta(D_s v)\, D_\alpha \phi\, dx = 0 \qquad \forall \phi \in H_0^1(\Omega) \tag{1.9}$$

and from De Giorgi-Nash's theorem we get

$$\int_{B_\rho} |\nabla^2 v|^2\, dx \le c\Big(\frac{\rho}{R}\Big)^{n-2+2\gamma} \int_{B_R} |\nabla^2 v|^2\, dx$$

for $\rho < R$ and for some positive γ; in particular we have that ∇v is Hölder-continuous.

Now we can rewrite (1.9) as

$$\int a_{\alpha p_\beta}((\nabla v)_R) D_\beta(D_s v) D_\alpha \phi + \int [a_{\alpha p_\beta}(\nabla v) - a_{\alpha p_\beta}((\nabla v)_R)] D_\beta(D_s v) D_\alpha \phi = 0$$

therefore, as we have done many times, aplitting v and using the L^p-estimate for $|D^2 v|$, we get

$$\int_{B_\rho(x_0)} |\nabla^2 v|^2 dx \leq c\left[\left(\frac{\rho}{R}\right)^n + \sigma\right] \int_{B_R(x_0)} |\nabla^2 v|^2 dx \tag{1.10}$$

and, since $a_{\alpha p_\beta}$ and ∇v are continuous, σ is small for small R, uniformly with respect to x_0. Using Lemma 2.1, Chapter III, we deduce from (1.10) the following estimate

$$\int_{B_\rho} |\nabla^2 u|^2 \leq c\left(\frac{\rho}{R}\right)^{n-\varepsilon} \int_{B_R} |\nabla^2 u|^2 dx \qquad \varepsilon > 0$$

which implies through a simple use of Caccioppoli's and Poincaré's estimates

$$\int_{B_\rho} |\nabla v - (\nabla v)_\rho|^2 dx \leq c\left(\frac{\rho}{R}\right)^{n+2-\varepsilon} \int_{B_R} |\nabla v - (\nabla v)_R|^2 dx\,. \tag{1.11}$$

The second remark is the following. Without loss in generality we may assume $a_\alpha(0) = 0$; then we have

$$\int\!\!\int_0^1 a_{\alpha p_\beta}(t\nabla v)\, dt\, D_\beta v D_\alpha \phi\, dx = 0 \qquad \forall \phi \in H_0^1(\Omega)$$

which can be rewritten as

$$\int\!\!\int_0^1 a_{\alpha p_\beta}(t(\nabla v)_R) D_\beta v D_\alpha \phi + \int\!\!\int_0^1 [a_{\alpha p_\beta}(t\nabla v) - a_{\alpha p_\beta}(t(\nabla v)_R)]\, dt D_\beta v D_\alpha \phi = 0$$
$$\forall \phi \in H_0^1\,.$$

Hence we deduce, as before,

$$\int_{B_\rho} |\nabla v|^2 dx \le c\left[\left(\frac{\rho}{R}\right)^n + \sigma\right] \int_{B_R} |\nabla v|^2 dx$$

and therefore

$$\int_{B_\rho} |\nabla v|^2 dx \le c\left(\frac{\rho}{R}\right)^{n-\varepsilon} \int_{B_R} |\nabla v|^2 dx \qquad \varepsilon > 0 . \tag{1.12}$$

Now we are ready for the proof. First assume that $a_\alpha = a_\alpha(x,p)$ with $a_\alpha(x,p)$ continuous in x. Let u be a bounded weak solution to

$$\int [a_\alpha(x,\nabla u) D_\alpha \phi + b(x,u,\nabla u)\phi] dx = 0 \qquad \forall \phi \in H_0^1 \cap L^\infty .$$

Split u as $u = v + (u-v)$ where v is the weak solution to

$$\begin{cases} \displaystyle\int_{B_R(x_0)} a_\alpha(x_0,\nabla v) D_\alpha \phi \, dx = 0 & \forall \phi \in H_0^1(B_R(x_0)) \\ u - v \in H_0^1(B_R(x_0)) & \end{cases}$$

then from (1.12) we get

$$\int_{B_\rho} |\nabla v|^2 dx \le c\left(\frac{\rho}{R}\right)^{n-\varepsilon} \int_{B_R} |\nabla v|^2 dx .$$

On the other hand

$$\int_{B_R} [a_\alpha(x_0,\nabla u) - a_\alpha(x_0,\nabla v)] D_\alpha \phi + \int_{B_R} [a_\alpha(x,\nabla u) - a_\alpha(x_0,\nabla u)] D_\alpha \phi +$$

$$+ \int_{B_R} b(x,u,\nabla u)\phi = 0 \qquad \forall \phi \in H_0^1 \cap L^\infty .$$

Therefore, since for $x \in B_R(x_0)$

$$|a_\alpha(x,\nabla u) - a_\alpha(x_0,\nabla u)| \leq \omega(R)\,|\nabla u|$$

and a_α is a monotone vector field with respect to p,[5)] we get

$$\int_{B_\rho} |\nabla u|^2 dx \leq c\left\{\left(\frac{\rho}{R}\right)^{n-2} + \left[\omega(R) + g\left(R^{2-n}\int_{B_R} |\nabla u|^2\right)\right]\right\} \int_{B_R} |\nabla u|^2 dx$$

which implies that $u \in C^{0,\gamma}$ for all $\gamma < 1 - \varepsilon/2$, through Lemma 2.1, Chapter III.

Therefore we can conclude that under the assumptions of Theorem 1.2 one has $u \in C^{0,\alpha}_{\mathrm{loc}}(\Omega)$ for all $\alpha < 1$. Now split u in $B_R(x_0) \subset\subset \Omega$ as $u = U + (u-U)$ where U is the weak solution to the Dirichlet problem

$$(1.13) \qquad \begin{cases} \displaystyle\int_{B_R(x_0)} a_\alpha(x_0, u_{x_0,R}, \nabla U) D_\alpha \phi = 0 & \forall \phi \in H_0^1(B_R(x_0)) \\ U - u \in H_0^1(B_R(x_0)) \end{cases}$$

then we have (estimate (1.11))

$$(1.14) \qquad \int_{B_\rho(x_0)} |\nabla U - (\nabla U)_\rho|^2 dx \leq c\left(\frac{\rho}{R}\right)^{n+2-\varepsilon} \int_{B_R(x_0)} |\nabla U - (\nabla U)_R|^2 dx$$

5) $[a_\alpha(x,p) - a_\alpha(x,q)](p_\alpha - q_\alpha) \geq \nu|p-q|^2$.

while for $w = u - U$ we have

$$\int_{B_R(x_0)} [a_\alpha(x_0, u_{x_0,R}, \nabla u) - a_\alpha(x_0, u_{x_0,R}, \nabla U)] D_\alpha \phi =$$

(1.15)
$$= \int_{B_R} \{[a_\alpha(x_0, u_{x_0,R}, \nabla u) - a_\alpha(x,u,\nabla u)] D_\alpha \phi - b(x,u,\nabla u)\phi\} dx$$

$\forall \phi \in H_0^1 \cap L^\infty$. Now note that $a_\alpha(x,u,p)$ is a monotone vector field in p and that there exists a nonnegative bounded increasing concave function $\omega(t)$, $\omega(0) = 0$, such that

$$\sum_\alpha |a_\alpha(x,u,p) - a_\alpha(y,v,p)|^2 \leq \omega(|x-y|^2, |u-v|^2)\,|p|^2$$

$$\omega(t) \leq \text{const}\ |t|^\delta$$

and finally note that, using the L^p-estimate and the boundedness of ω, we have

$$\sum_\alpha \int_{BO} |a_\alpha(x_0, u_{x_0,R}, \nabla u) - a_\alpha(x,u,\nabla u)|^2 dx \leq$$

$$\leq \int \omega(|x-x_0|^2 + |u-u_{x_0,R}|^2)\,|\nabla u|^2 dx \leq cR^{2\delta\alpha} R^{n-2+2\alpha}$$

$$\int_{B_R} (1+|\nabla u|^2)\,|u-U| \leq c\left(\int (1+|\nabla u|^2)^{q/2} dx\right)^{2/q} \left(\int |u-U|^2 dx\right)^{1-2/q} \leq$$

$$\leq c \int_{B_{2R}} (1+|\nabla u|^2) dx \left(R^{2-n} \int_{B_R} |\nabla u|^2\right)^{1-2/q} \leq \text{const}\ R^{n-2+2\alpha+2\alpha(1-2/q)}\,.$$

Therefore from (1.15), inserting $\phi = u - U$, we conclude

$$\int_{B_R} |\nabla(u-U)|^2 dx \le \text{const } R^{n-2+2\alpha(1+\delta \wedge (1-2/q))} \tag{1.16}$$

for some $q > 2$ near to 2. If α is chosen so close to 1 that $2\alpha(1+\delta \wedge (1-2/q)) > 2$ we conclude from (1.14), (1.16) that ∇u is Hölder-continuous, and in particular that ∇u is locally bounded. But then we have

$$\int_{B_R} |\nabla(u-U)|^2 dx \le c\left[R^{n+2\delta} + \int_{B_R} |u-U|\, dx\right] .$$

The last integral is easily estimated

$$\int_{B_R} |u-U|\, dx \le R^{\frac{n+2}{2}} \left[\int_{B_R} |\nabla(u-U)|^2 dx\right]^{1/2}$$

therefore

$$\int_{B_R} |\nabla(u-U)|^2 dx \le \text{const } R^{n+2\delta}$$

which again together with (1.14) finally implies $u \in C^{1,\delta}_{\text{loc}}(\Omega)$. q.e.d.

2. *Minima of variational integrals*: $N = 1$

Let us consider the functional

$$J[u;\Omega] = \int_{\Omega} F(x,u,\nabla u)\, dx \qquad F(x,u,p) : \Omega \times \mathbf{R} \times \mathbf{R}^n \to \mathbf{R}$$

where

i) F is a Caratheodory function, i.e. F is measurable in x for all (u,p) and continuous in (u,p) for almost all x

ii) there exist positive constants a and k such that

$$|p|^2 - k \leq F(x,u,p) \leq a|p|^2 + k\,. \tag{2.1}$$

In [191] the boundedness of a function is proved, which minimizes $J[u;\Omega]$ among all functions taking prescribed values $z(x)$ on $\partial\Omega$, provided $z(x)$ is bounded. Therefore, requiring some smoothness of the function F, suitable growth conditions for its partial derivatives F_u and F_p and the ellipticity condition (i.e. convexity of F with respect to p) Theorems 1.1 and 1.2 of Section 1 are applicable and we get regularity of the minimum points.

Of course, some smoothness of F is necessary if one wants to prove the differentiability of the minima; on the other hand, if we look only for the continuity of the solution, such assumptions seem superfluous, compare with Morrey's result Section 3, Chapter V, and it would be preferable to derive it directly from the minimizing property of u .

In this section we shall investigate the Hölder-continuity of the minima, working directly with the functional J instead of working with its Euler equation. The results we shall present here are due to M. Giaquinta and E. Giusti [114].

Local boundedness. We shall assume slightly more general hypotheses on F than those previously described. More precisely

i) F *is a Caratheodory function*

ii) *There exist positive constants* a *and* b *and a real number* $m > 1$ *such that*

$$|p|^m - b(|u|^\alpha + 1) \leq F(x,u,p) \leq a|p|^m + b(|u|^\alpha + 1) \tag{2.2}$$

where $m \leq \alpha < m^* = \frac{mn}{n-m}$.[6)]

---

6)We shall restrict ourselves to the case $1 < m < n$. When $m > n$ every function in $H^{1,m}$ is trivially Hölder-continuous; and we have considered the case $m = n$ in Section 3, Chapter V.

We shall consider *local minimum points for the functional* $J[u;\Omega]$, i.e. *functions* $u \in H^{1,m}_{loc}(\Omega)$ *such that for every* $\phi \in H^{1,m}(\Omega)$ *with* $\operatorname{spt}\phi \subset\subset \Omega$ *we have*

$$J[u; \operatorname{spt}\phi] \leq [u+\phi; \operatorname{spt}\phi] .$$

In [191] the boundedness of a global minimum point taking prescribed value $z(x)$ at $\partial\Omega$ is proved provided $z(x)$ is bounded: here we shall consider the problem of the local boundedness of minima independently of the boundary data.

THEOREM 2.1. *Let (2.2) hold, and let* $u \in H^{1,m}_{loc}(\Omega)$ *be a local minimum for the functional* J. *Then* u *is locally bounded in* Ω.

Proof. We may suppose Ω bounded and $u \in H^{1,m}(\Omega)$. Let $x_0 \in \Omega$, and denote by B_s the ball $B_s(x_0)$. For $k > 0$ set

$$A_k = \{x \in \Omega : u(x) > k\} \qquad A_{k,s} = A_k \cap B_s . \tag{2.3}$$

Let $w = \max(u-k, 0)$ and let $\eta(x)$ be a C^∞ function with $\operatorname{spt}\eta \subset B_s$, $0 \leq \eta \leq 1$ $\eta \equiv 1$ on B_t, $|\nabla\eta| \leq 2(s-t)^{-1}$. If $v = u - \eta w$, we have, using the minimality of u and (2.2):

$$\int_{A_{k,s}} |\nabla u|^m dx \leq \gamma_1 \left\{ \int_{A_{k,s}} (1-\eta)^m |\nabla u|^m dx + \int_{A_{k,s}} w^m |\nabla\eta|^m dx + \right.$$

$$\left. + \int_{A_{k,s}} w^\alpha dx + (1+k^\alpha)\,|A_{k,s}| \right\} .$$

Now we observe that if $w \in H^{1,m}(B_s)$ and $|\operatorname{spt} w| \leq \frac{1}{2}|B_s|$ we have (compare with footnote 13, Chapter V) the Sobolev inequality

$$\left(\int_{B_s} w^{m^*}\,dx\right)^{m/m^*} \leq c_1(m,n)\int_{B_s}|\nabla w|^m\,dx \tag{2.5}$$

and therefore, if $m \leq \alpha < m^*$:

$$\int_{B_s} w^{\alpha}\,dx \leq \|w\|_{m^*}^{\alpha-m}|B_s|^{1-\alpha/m^*}\left(\int_{B_s} w^{m^*}\,dx\right)^{m/m^*} \leq$$

$$\leq c_1\|w\|_{m^*}^{\alpha-m}|B_s|^{1-\alpha/m^*}\int_{B_s}|\nabla w|^m\,dx\,. \tag{2.6}$$

We can choose T so small that for $s \leq T$ we get

$$c_1\|u\|_{m^*}^{\alpha-m}|B_s|^{1-\alpha/m^*} \leq \frac{1}{2\gamma_1}\,. \tag{2.7}$$

On the other hand, we have

$$k^{m^*}|A_k| \leq \|u\|_{m^*}^{m^*}$$

and therefore for $k \geq k_0$ we have

$$|A_k| < \frac{1}{2}|B_{T/2}|\,.$$

For such values of k we then have $|\text{spt } w| < \frac{1}{2}|B_{T/2}|$ and therefore, if $T/2 \leq s \leq T$:

$$\int_{B_s} w^{\alpha}\,dx \leq \frac{1}{2\gamma_1}\int_{A_{k,s}}|\nabla u|^m\,dx \tag{2.8}$$

since $\|u\|_{m^*} \geq \|w\|_{m^*}$.

In conclusion, if $\frac{T}{2} \leq t < s \leq T$, we have from (2.4), (2.8):

$$\int_{A_{k,s}} |\nabla u|^m dx \leq 2\gamma_1 \left\{ \int_{A_{k,s} \setminus A_{k,t}} |\nabla u|^m dx + \frac{1}{(s-t)^m} \int_{A_{k,s}} w^m dx + \right. \\ \left. + (1+k^\alpha)\, |A_{k,s}| \right\} .$$

Suppose now $\frac{T}{2} \leq \rho \leq t < s \leq R \leq T$; we get

$$\int_{A_{k,t}} |\nabla u|^m dx \leq 2\gamma_1 \int_{A_{k,s} \setminus A_{k,t}} |\nabla u|^m dx + 2\gamma_1 \left\{ (s-t)^{-m} \int_{A_{k,R}} w^m dx + \right. \\ \left. + (1+k^\alpha)\, |A_{k,R}| \right\} .$$

We now proceed as in the proof of Theorem 3.1, Chapter V. Adding to both sides $2\gamma_1$ times the left-hand side we get

$$\int_{A_{k,t}} |\nabla u|^m dx \leq \frac{2\gamma_1}{2\gamma_1 + 1} \int_{A_{k,s}} |\nabla u|^m dx + \left\{ (s-t)^{-m} \int_{A_{k,R}} w^m dx + \right. \\ \left. + (1+k^\alpha)\, |A_{k,R}| \right\} .$$

Then we apply Lemma 3.1, Chapter V and conclude that

$$\int_{A_{k,\rho}} |\nabla u|^m dx \leq \gamma_2 \left\{ (R-\rho)^{-m} \int_{A_{k,R}} w^m dx + (1+k^\alpha)|A_{k,R}| \right\} . \tag{2.9}$$

Finally we estimate

$$(1+k^{\alpha})|A_{k,R}| \le 2k^{\alpha}|A_{k,R}| \le 2(k^{m^*}|A_{k,R}|)^{\frac{\alpha-m}{m^*}} k^m |A_{k,R}|^{1-\frac{\alpha-m}{m^*}} \le$$

$$\le 2\|u\|_{m^*}^{\alpha-m} k^m |A_{k,R}|^{1-\frac{m}{n}+\left(1-\frac{\alpha}{m^*}\right)} .$$

Introducing the last expression into (2.9) we get for $\frac{T}{2} \le \rho < R \le T$:

$$(2.10) \qquad \int_{A_{k,\rho}} |\nabla u|^m \, dx \le \gamma_3 \left\{ (R-\rho)^{-m} \int_{A_{k,R}} (u-k)^m \, dx + k^m |A_{k,R}|^{1-\frac{m}{n}+\left(1-\frac{\alpha}{m^*}\right)} \right\}$$

Since $-u$ minimizes the functional

$$\bar{J}[v;\Omega] = \int_{\Omega} \bar{F}(x,v,\nabla v)\, dx$$

with $\bar{F}(x,v,p) = F(x,-v,-p)$ satisfying the same growth condition (2.3), inequality (2.10) holds with u replaced by $-u$. We may then apply to both u and $-u$ Lemma 5.4 of Chapter II of [191][7] and conclude that u is bounded in $B_{T/2}$. q.e.d.

---

[7] LEMMA. *Suppose that* $B_{\rho_0}(x_0) \subset \Omega$. *Let us suppose that, for arbitrary* κ *greater than some* $\hat{k}$ *and for arbitrary spheres* $B_\rho(x_0)$ *and* $B_{\rho-\sigma\rho}(x_0)$, *where* $\rho_0 - \sigma\rho_0 \le \rho - \sigma\rho < \rho \le \rho_0$, *the function* $u(x)$ *satisfies the inequalities*

$$\int_{A_{k,\rho-\sigma\rho}} |\nabla u|^m \, dx \le \gamma[(\sigma\rho)^{-m} \int_{A_{k,\rho}} (u-k)^m \, dx + \rho^{\varepsilon-m} k^{\alpha} |A_{k,\rho}|^{1-\frac{m}{n}+\varepsilon}$$

where σ_0, γ, α *and* ε *are positive constants and where* $\tau_0 < 1$, $\varepsilon \le \frac{m}{n}$, *and* $m \le \alpha < \varepsilon m + m$. *Then, in the sphere* $B_{\rho_0 - \tau\rho_0}$, *the quantity* $\operatorname{essmax} u(x)$ *is bounded above by a constant depending only on* $\sigma_0, \hat{k}, n, m, \gamma, \varepsilon, \alpha$ *and the quantity*

$$\rho_0^{-n} \int_{A_{\hat{k},\rho_0}} (u(x)-\hat{k})^m \, dx .$$

The above result may easily be generalized; for instance one might assume that the constant b appearing in (2.2) is actually a function belonging to some suitable L^η space. Moreover, one can assume that the minimizing function $u(x)$ belongs to $H^{1,m} \cap L^q(\Omega)$, for some $q \geq m^*$. In this case the conclusion of the theorem holds if $\eta > \frac{n}{m}$ and $a < m\frac{n+q}{n} - \frac{q}{\eta}$. But we do not insist on this point.

Hölder continuity. An argument similar to the one above will give now the Hölder-continuity of the local minima for the functional $J[u;\Omega]$. We suppose that the function $F(x,u,p)$ satisfies the growth condition

$$|p|^m - b(M) \leq F(x,u,p) \leq a(M)\,|p|^m + b(M) \tag{2.11}$$

for every $x \in \Omega$, $|u| \leq M$ and $p \in R^n$.

We would like to point out explicitly that we are not assuming differentiability or convexity of F in p.

THEOREM 2.2. *Let (2.11) hold and let* $u(x)$ *be a function in* $H^{1,m}_{loc}(\Omega) \cap L^\infty_{loc}(\Omega)$ *minimizing the functional* $J[u;\Omega]$. *Then* u *is Hölder-continuous in* Ω.

Proof. We take as before $v = u - \eta w$; from $J[u] \leq J[v]$ we easily deduce, using (2.11) with $M = \sup |u|$, $\gamma_4 = \gamma_4(M)$

$$\int_{A_{k,s}} |\nabla u|^m dx \leq \gamma_4 \left\{ \int_{A_{k,s}} (1-\eta)^m |\nabla u|^m dx + \int_{A_{k,s}} w^m |\nabla \eta|^m dx + |A_{k,s}| \right\}.$$

Observing that $\eta \equiv 1$ on B_t and that $|\nabla \eta| \leq 2(s-t)^{-1}$ we get therefore for $R \geq s > t$:

$$\int_{A_{k,t}} |\nabla u|^m dx \leq \gamma_5 \left\{ \int_{A_{k,s}\setminus A_{k,t}} |\nabla u|^m dx + (s-t)^{-m} \int_{A_{k,R}} (u-k)^m + |A_{k,R}| \right\}.$$

Arguing again as in the above, we conclude using Lemma 3.1, Chapter V:

$$(2.12) \qquad \int_{A_{k,\rho}} |\nabla u|^m dx \leq \gamma_6 \left\{ (R-\rho)^{-m} \int_{A_{k,R}} (u-k)^m dx + |A_{k,R}| \right\} .$$

The same inequality holds with u replaced by $-u$, and therefore the function u belongs to the De Giorgi class $\mathcal{B}_m(\Omega, M, \gamma_6, 1, 0)$ [8] of [191]. Applying Theorem 6.1 of Chapter II of [191][9] we conclude that u is locally Hölder-continuous in Ω. q.e.d.

Using the same argument it is not difficult to prove regularity up to the boundary for solutions of the Dirichlet problem, provided the boundary datum is itself Hölder-continuous on $\partial\Omega$ and $\partial\Omega$ is sufficiently smooth. In fact, inequality (2.12) still holds when the ball B_R intersects $\partial\Omega$, provided the constant k is greater than $\sup_{\partial\Omega \cap B_R} u$, so that we can apply the result of [191], Chapter II.7.

For the sake of completeness we mention that a result of the type of Theorem 2.2 has been proved by J. Frehse [93], under very strong assumptions on F.

---

[8] *De Giorgi's classes.* $\mathcal{B}_m\left(\Omega, M, \gamma, \delta, \frac{1}{q}\right)$ denotes the class of functions $u(x)$ in $H^{1,m}(\Omega)$ with $\max_{\Omega} |u| \leq M$ such that for u and $-u$ the following inequalities are valid in an arbitrary sphere $B_\rho \subset \Omega$ for arbitrary σ in $(0,1)$:

$$\int_{A_{k,\rho-\sigma\rho}} |\nabla u|^m dx \leq \gamma \left[\frac{1}{\sigma^m \rho^m \left(1-\frac{n}{q}\right)} \max_{A_{k,\rho}} [u(x)-k]^m + 1 \right] |A_{k,\rho}|^{1-\frac{m}{q}}$$

for $k \geq \max_{B_\rho} u - \delta$.

[9] THEOREM. *Let* $u(x)$ *be an arbitrary function in* $\mathcal{B}_m\left(\Omega, M, \gamma, \delta, \frac{1}{q}\right)$ *and* $B_{\rho_0}(x_0) \subset \Omega$ $\rho_0 \leq 1$. *Then for an arbitrary* $B_\rho(x_0)$ $\rho < \rho_0$, *the oscillation of* $u(x)$ *in* B_ρ *satisfies the inequality*

$$\operatorname*{osc}_{B_\rho} u \leq c\left(\frac{\rho}{\rho_0}\right)^{\alpha}$$

for some positive α.

3. *Systems of diagonal form*

Let us consider systems of partial differential equations of the following form

$$(3.1) \qquad -D_\alpha(a^{\alpha\beta}(x)D_\alpha u^i) = f_i(x,u,\nabla u) \qquad i = 1,\cdots,N$$

under the following assumptions:

i) $a^{\alpha\beta}(x) \in L^\infty(\Omega)$ *and*

$$|a^{\alpha\beta}| \le \mu$$

$$a^{\alpha\beta}\xi_\alpha\xi_\beta \ge \lambda|\xi|^2 \qquad \forall \xi \in \mathbf{R}^n;\ \lambda > 0$$

ii) $f = (f_1,\cdots,f_N)$ *is a Caratheodory function satisfying the growth condition*

$$|f(x,u,p)| \le a|p|^2 + b$$

with a, b *constants.*

In recent years, considerable attention has been directed towards systems of type (3.1) because of their relevance in problems arising in differential geometry such as harmonic mappings of manifolds or surfaces of prescribed mean curvature.

The reader may refer to [160] (see also [158][159][161]) for a survey on results and proofs, including also an account of the history of the problem and its connection with the problem of harmonic maps of Riemannian manifolds.

The following regularity theorem due to S. Hildebrandt and K.-O. Widman [165] and M. Wiegner [302][303] can be considered, in view of examples 3.6, Chapter II and 2.1, Chapter V, as an optimal result (see [118] for a simpler proof).

THEOREM 3.1. *Every weak solution* u *of (3.1) with* $\sup_\Omega |u(x)| \le M$ *is locally Hölder-continuous in* Ω, *provided* $aM < \lambda$. *For every subdomain* $\Omega' \subset\subset \Omega$, *the Hölder norm of* u *restricted to* Ω' *can be estimated in terms of* $u, \lambda, \mu, a, b, M, \operatorname{dist}(\Omega', \partial\Omega)$.

In connection with the study of diagonal systems we would like to mention also [164][304][299], [176][177] for uniqueness results, [298] for the existence. Moreover we would like to mention [167][209][173][212] where Liouville type theorems are proved.

We refer to the papers quoted above for the proof of Theorem 3.1; here we only point out that under the stronger assumption

$$2aM < \lambda . \tag{3.2}$$

Theorem 3.1 can be proved as Theorem 1.1. Let us sketch the proof.

(a) Because of (3.2) we know that the L^p-estimate for the gradient of a weak solution of (3.1) is true.

(b) Splitting u in B_R as $u = v+(u-v)$ where v is the weak solution to

$$\begin{cases} \displaystyle\int_{B_R} a^{\alpha\beta} D_\alpha v^i D_\beta \phi^i \, dx = 0 & \forall \phi \in H_0^1(B_R) \\ v - u \in H_0^1(B_R) \end{cases}$$

from De Giorgi-Nash theorem we get for $\rho < R$

$$\int_{B_\rho} |\nabla v|^2 dx \leq c\left(\frac{\rho}{R}\right)^{n-2+2\alpha} \int_{B_R} |\nabla v|^2 dx$$

for some positive α; so that estimating $u-v$ in the usual way (using step (a) compare with Section 1, Chapter VI) we obtain: u is Hölder-continuous in an open set Ω_0 and

$$\Omega \setminus \Omega_0 \subset \left\{ x_0 \in \Omega : \liminf_{R \to 0^+} R^{2-n} \int_{B_R(x_0)} |\nabla u|^2 dx > \varepsilon_0 \right\}$$

where ε_0 is a positive constant.

(c) Therefore in order to prove the regularity result it remains to show that for all $x_0 \in \Omega$ there exists R such that

$$R^{2-n} \int_{B_R(x_0)} |\nabla u|^2 \leq \varepsilon_0 .$$

This can be done exactly as in Section 1, step III of the proof of Theorem 1.1 choosing $\phi = \eta u^2$ as test function in the weak formulation

$$\int_\Omega a^{\alpha\beta} D_\alpha u^i D_\beta \phi^i \, dx = \int f_i(x,u,\nabla u)\, \phi^i dx \qquad \forall \phi \in H_0^1 \cap L^\infty(\Omega, R^N) .$$

In fact we get

$$(\lambda - aM) \int |\nabla u|^2 \eta \, dx \leq -\frac{1}{2} \int a^{\alpha\beta} D_\alpha |u|^2 D_\beta \, \eta \, dx + b \int \eta \, dx \qquad \forall \eta \in H_0^1(\Omega), \eta \geq 0 .$$

The same scheme of proof can be used in different situations. For example assume that

$$f_i(x,u,\nabla u) = \sum_{\alpha=1}^{n} g_\alpha(x,u,\nabla u) D_\alpha u^i \tag{3.3}$$

where

$$|g_\alpha(x,u,\nabla u)| \leq L|\nabla u| + N$$

then we have

PROPOSITION 3.1. *Assume (3.3). Then every weak solution to (3.1) with* $\sup |u| \leq M < +\infty$ *is locally Hölder-continuous in* Ω *and the a priori estimate holds.*

Proof. Inserting as test function $(u - u_R) e^{t|u-u_R|^2} \eta^2$ we deduce, compare with Proposition 2.1, Chapter V, an L^p-estimate for the gradient.

Therefore (a) and (b) above hold; (c) can be simply achieved inserting $ue^{t|u|^2}\eta$ as test function. q.e.d.

Let us remark that condition (3.2) is involved in the proof only in step (a); on the other hand note that an L^p-estimate is implied by Theorem 3.1.

One-sided condition on f. In [164] S. Hildebrandt and K.-O. Widman, in connection with the study of harmonic maps of manifolds, conjectured that every bounded solution of (3.1) were Hölder-continuous if the right-hand side f satisfied the inequalities

$$(3.4)\qquad \begin{cases} |f(x,u,p)| \leq a|p|^2 + b \\ u^i f_i(x,u,p) \leq \lambda^*|p|^2 + b^* \qquad \lambda^* < \lambda . \end{cases}$$

On the basis of a counterexample to the Liouville theorem due to M. Meier [209], P. A. Ivert [173] has shown that in general a priori estimates of the Hölder norm of solutions u of (3.1) cannot exist whenever (3.4) holds; and M. Struwe [288] has shown an example of a singular solution in dimension $n \geq 3$. In dimension $n = 2$, instead, J. Frehse [95] has proved the existence of a smooth solution and M. Wiegner [306] that all solutions are smooth in the interior (compare also with [211]).

We shall prove now the Hölder-continuity of solutions u provided (3.1) is the 'Euler equation of a functional' and u a minimum point, compare with [114], which is essentially the case of harmonic maps of manifold considered in [163], see [76] [77] [75] for information on harmonic mappings of manifolds.

Let us consider the functional

$$(3.5)\qquad \int_\Omega g_{ij}(u)A^{\alpha\beta}(x)D_\alpha u^i D_\beta u^j\, dx$$

where $A^{\alpha\beta}$ and g_{ij} are definite positive symmetric matrices; and let

$u \in H^1_{loc}(\Omega, R^N)$ be a local minimum point. From Section 3, Chapter VI, we know that $u \in C^{0,\alpha}(\Omega_0)$ and

$$x_0 \in \Omega\backslash\Omega_0 \iff \liminf_{R\to 0^+} R^{2-n} \int_{B_R(x_0)} |\nabla u|^2 dx > \varepsilon_0 .$$

Of course we are assuming g_{ij} continuous (for the sake of simplicity uniformly continuous) and $A^{\alpha\beta}$ continuous (but it would be sufficient to assume $A^{\alpha\beta} \in L^\infty$).

Let now u be a bounded minimum, then u satisfies[10)]

$$\int A^{\alpha\beta} g_{ij} D_\alpha u^i D_\beta \phi^j dx + \int A^{\alpha\beta} g_{ij,u^l} D_\alpha u^i D_\beta u^j \phi^l dx = 0 \quad \forall \phi \in H^1_0 \cap L^\infty(\Omega, R^N)$$

which, denoting by (g^{ij}) the inverse matrix of g, can be rewritten as

$$\int A^{\alpha\beta} D_\alpha u^i D_\beta \eta^i dx + \int A^{\alpha\beta}(g_{hk,u^l} - g_{hl,u^k}) D_\alpha u^h D_\beta u^k g^{li} \eta^i dx = 0$$

for all $\eta \in H^1_0 \cap L^\infty(\Omega; R^N)$, i.e. as system (3.1) with

$$f_i(x,u,\nabla u) = -A^{\alpha\beta}(g_{hk,u^l} - g_{hl,u^k}) D_\alpha u^h D_\beta u^k g^{li} .$$

Therefore we immediately see that assuming (3.4) we can carry on step (b), i.e. the analogous of step III in the proof of Theorem 1.1, and show that $\Omega\backslash\Omega_0 = \emptyset$.

Let us remark that if u is bounded on $\partial\Omega$, since it essentially happens that $|u|^2$ is a subsolution for an elliptic operator, then u is bounded in Ω.

---

10) Now we are assuming that g_{ij} be differentiable.

Therefore we can conclude

THEOREM 3.2. *Let* $\{A^{\alpha\beta}(x)\}$ *and* $\{g_{ij}(u)\}$ *be smooth symmetric definite positive matrices with*

$$A^{\alpha\beta}(x)\,\xi_\alpha\xi_\beta \geq \lambda|\xi|^2 \,.$$

Assume that

$$-A^{\alpha\beta}(g_{hk,u^l} - g_{hl,u^k})D_\alpha u^h D_\beta u^k g^{li} u^i \leq \lambda^*|u|^2$$

where u *is a bounded minimum point for the functional (3.5). Then* u *is locally Hölder-continuous in* Ω.

Higher regularity then follows in the usual way.

Besides the papers we have already quoted, we refer also to [119], [291] for the use of similar ideas still in connection with the regularity of harmonic mappings of manifolds.

4. *Functionals depending on the modulus of the gradient*

Let us consider the regular functional of the calculus of variations

$$\int_\Omega F(\nabla u)\,dx \tag{4.1}$$

under the standard hypothesis

$$\begin{cases} \lambda|p|^2 \leq F(p) \leq \mu|p|^2 + \gamma \\ |f_p| \leq K|p| \qquad |F_{pp}| \leq 2k \\ F_{p^i_\alpha p^j_\beta}\,\xi^i_\alpha\xi^j_\beta \geq 2\nu|\xi|^2 \qquad\qquad \forall\xi;\ \nu>0 \end{cases} \tag{4.2}$$

and let us restrict our consideration to functionals of the type (4.1), (4.2) which depend only on the modulus of the gradient, i.e.

$$F(\nabla u) = g(|\nabla u|^2) . \tag{4.3}$$

Then (4.2) implies

$$\text{(4.2)}' \quad \begin{cases} \lambda|Q| \le g(Q) \le \mu|Q| + \gamma \\ |g'| \le k \\ [g'(|\nabla u|^2)\delta_{ij}\delta_{\alpha\beta} + 2g''(|\nabla u|^2)D_\alpha u^i D_\beta u^j]\xi^i_\alpha \xi^j_\beta = A^{\alpha\beta}_{ij}\xi^i_\alpha\xi^j_\beta \ge \nu|\xi|^2 \\ |A^{\alpha\beta}_{ij}| \le k . \end{cases}$$

Note moreover that

$$0 < \sigma_1 \le g'(|\nabla u|^2) \le \sigma_2 \qquad \sigma_1, \sigma_2 \in R .$$

Then from the Euler equation

$$\int_\Omega g'(|\nabla u|^2) D_\alpha u^i D_\alpha \phi^i \, dx = 0 \qquad \forall \phi \in H^1_0(\Omega, R^N)$$

we get immediately, through De Giorgi-Nash's theorem, that u is locally Hölder-continuous. Moreover writing the equation in variation we obtain that the derivatives of u are Hölder-continuous in an open set Ω_0 and that there exists $\varepsilon_0 > 0$ such that

$$\Omega \backslash \Omega_0 \subset \left\{ x_0 \in \Omega : R^{2-n} \int_{B_R(x_0)} |\nabla^2 u|^2 dx > \varepsilon_0 \qquad \forall R < R_0 \right\} .$$

Then we have

THEOREM 4.1. *The weak stationary points of functional (4.1), (4.3) are* C^1*-Hölder-continuous in* Ω.

Proof. In order to prove the theorem, it is sufficient to show that for all $x_0 \in \Omega$ there exists R such that

$$R^{2-n} \int_{B_R(x_0)} |\nabla^2 u|^2 dx < \varepsilon_0 . \tag{4.4}$$

Now the equation in variation is

$$D_\alpha[g' D_\alpha D_s u^i + g'' D_s |\nabla u|^2 D_\alpha u^i] = 0 \qquad i = 1, \cdots, N \quad s = 1, \cdots, n .$$

Multiplying by $D_s u^i$ and summing on s and i we get

$$0 = D_\alpha[g' D_\alpha D_s u^i D_s u^i + g'' D_\alpha u^i D_s u^i D_s |\nabla u|^2] - g' D_\alpha D_s u^i D_\alpha D_s u^i -$$
$$- 2g'' D_\alpha u^i D_\beta u^j D_\alpha D_s u^i D_\beta D_s u^j$$

which can be rewritten as

$$D_\alpha[A^{\alpha\beta} D_s |\nabla u|^2] = 2A^{\alpha\beta}_{ij} D_\alpha D_s u^i D_\beta D_s u^j$$

where

$$A^{\alpha\beta} = g'(|\nabla u|^2)\delta_{\alpha\beta} + 2g''(|\nabla u|^2) D_\alpha u^i D_\beta u^i$$

$A^{\alpha\beta}_{ij}$ is defined in (4.2)′ .

Now note that

$$A^{\alpha\beta}\xi_\alpha \xi_\beta \geq \nu|\xi|^2 \qquad \forall \xi > 0 ; \ \nu > 0$$

$$A^{\alpha\beta}_{ij} D_\alpha D_s u^i D_\beta D_s u^j \geq \nu |\nabla^2 u|^2 .$$

Therefore we get

$$D_\alpha[A^{\alpha\beta} D_s |\nabla u|^2] \geq \nu |\nabla^2 u|^2 \tag{4.5}$$

i.e. $|\nabla u|^2$ is a subsolution for an elliptic operator, and the required condition (4.4) follows exactly as in Section 1, step III of the proof of Theorem 1.1, inserting as test function in the weak formulation of (4.5) $|\nabla u|^2 \eta$. q.e.d.

Theorem 4.1 is a special case of a more general theorem due to K. Uhlenbeck [296], which refers essentially to solutions of Euler type systems of functionals of the type

$$\int (c+|\nabla u|^2)^{k/s}\, dx\,, \quad \int F(c+|\nabla u|^2)\, dx \qquad k \geq 2$$

where even $c = 0$ is allowed. We refer to [296] for the statement and the proof; and to [171][172] for some extensions (see also [83] for a different proof).

Chapter VIII
A FEW REMARKS AND EXTENSIONS

In this chapter we want to mention some extensions of the methods and results already described and to hint at the problem of the regularity up to the boundary.

1. *A few extensions*

The obstacle problem. Let u be a solution of the variational problem

$$\int_\Omega |\nabla u|^2 dx \to \min$$

$$u = 0 \text{ on } \partial\Omega; \quad u \geq \Psi \text{ in } \Omega$$

(of course we assume that $\Psi \leq 0$ on $\partial\Omega$), or let u satisfy the variational inequality (compare Chapter I):

$$(1.1) \qquad \int_\Omega \nabla u \nabla(u-v)\, dx \leq 0 \qquad \forall v \in H_0^1(\Omega) \qquad v \geq \Psi \text{ in } \Omega .$$

We now want to show that the method of Chapter III permits to prove the continuity of u, see [108][109]; i.e., roughly speaking we want to show that the regularity follows by comparing, locally on balls B_R, the weak solution u to (1.1) with the harmonic function U with boundary value (on ∂B_R) equal to u.

Let us split u as $u = U + (u-U)$ on $B_R(x_0) \subset \Omega$, where $U \in H^1(B_R(x_0))$ is the harmonic function in $B_R(x_0)$ with $U = u$ on ∂B_R, then we have for $\rho < R$ (see Chapter III)

$$(1.2)\qquad \int_{B_\rho(x_0)} |\nabla U|^2\,dx \le \tilde{c}_1\left(\frac{\rho}{R}\right)^n \int_{B_R(x_0)} |\nabla U|^2\,dx \le c_1\left(\frac{\rho}{R}\right)^n \int_{B_R(x_0)} |\nabla u|^2\,dx$$

while $u - U$ satisfies

$$(1.3)\qquad \int_{B_R(x_0)} \nabla(u-U)\,\nabla(u-v)\,dx \le 0 \quad \forall v \in H_0^1(B_R(x_0)) \;\; v \ge \Psi \;\text{ in }\; B_R(x_0)\,.$$

Now writing in (1.3) $u - U + U - v$ instead of $u - v$ and choosing, as it can be done

$$v = \max(U, \Psi) = U \vee \Psi$$

we easily get

$$(1.4)\qquad \int_{B_R(x_0)} |\nabla(u-U)|^2\,dx \le c_2 \int_{B_R(x_0)} |\nabla(U - U\vee\Psi)|^2\,dx\,.$$

But $U - U\vee\Psi \in H_0^1(B_R(x_0))$ satisfies

$$\int_{B_R(x_0)} \nabla(U - U\vee\Psi)\nabla\phi = -\int \nabla U\vee\Psi\,\nabla\phi \qquad \forall\phi \in H_0^1(B_R(x_0))$$

hence choosing $\phi = U - U\vee\Psi$, since $U\vee\Psi = \Psi$ for $x \in \operatorname{spt}(U - U\vee\Psi)$ we obtain

$$(1.5)\qquad \int_{B_R(x_0)} |\nabla(U - U\vee\Psi)|^2 \le c_3 \int_{B_R(x_0)} |\nabla\Psi|^2\,dx\,.$$

Therefore, assuming that

$$\int_{B_R} |\nabla \Psi|^2 dx \leq c_4 R^{n-2+2\alpha}$$

from (1.3), (1.5) we deduce

$$\int_{B_\rho(x_0)} |\nabla u|^2 dx \leq c_5 \left(\frac{\rho}{R}\right)^n \int_{B_R(x_0)} |\nabla u|^2 dx + c_6 R^{n-2+2\alpha}$$

which, through Lemma 2.1, Chapter III implies

$$\int_{B_\rho(x_0)} |\nabla u|^2 dx \leq c_7 \rho^{n-2+2\alpha}$$

i.e. $u \in C^{0,\alpha}_{loc}(\Omega)$. Concluding

THEOREM 1.1. *Let* u *be a weak solution to the variational inequality (1.1) and let the obstacle be such that* $|\nabla \Psi| \in L^{2,n-2+2\alpha}_{loc}(\Omega)$. *Then* $u \in C^{0,\alpha}_{loc}(\Omega)$.

By a simple use of the maximum principle, Theorem 1.1 can be proved under the weaker assumption that Ψ be only Hölder-continuous, see [109]. Moreover the above proof can be carried on for more general second order variational inequalities, and using the same ideas in Chapter VI one can study the Hölder- and $C^{1,\alpha}$ continuity for solutions of general second order nonlinear variational inequalities, see [108][109].

But we shall not insist and simply point out that although some result can be obtained also in the vector valued case, [108], in general the problem of the regularity for solutions of variational inequalities with full system operators is greatly open. We must mention, anyway, that precise results have been obtained in the vector valued case for diagonal variational inequalities, see e.g. [166]. Finally we would like to mention that very precise regularity results for variational inequalities with irregular obstacle have been obtained by J. Frehse and U. Mosco [97][98].

Systems of the type of the stationary Navier-Stokes system. Here we only want to mention that the *methods* and *results* we have presented for systems of the type

$$\int A_i^{\alpha}(x,u,\nabla u)D_{\alpha}\phi^i = \int B_i(x,u,\nabla u)\,\phi^i\,dx \qquad \forall \phi \in C_0^{\infty}(\Omega, R^N)$$

in the different situations, can be carried on to nonlinear systems of the type of the stationary Navier-Stokes system, i.e. to systems of the type

$$\operatorname{div} u = g$$

(1.5)
$$\int_{\Omega} A_i^{\alpha}(x,u,\nabla u)D_{\alpha}\phi^i = \int_{\Omega} B_i(x,u,\nabla u)\,\phi^i$$

for all solenoidal vector field ϕ with compact support, i.e. $\phi \in H_0^1(\Omega,R^n)$ with $\operatorname{div}\phi = 0$ in Ω, see [123]. There it is assumed that

(1.6)
$$\begin{cases} |A_i^{\alpha}(x,u,p)| \leq L(|p| + |u|^{r/2}) + f_1(x) \\ |B_i(x,u,p)| \leq L(|p|^{r_1} + |u|^{r_0}) + f_0(x) \\ r \leq 2^*, \quad r_1 \leq 2/2^*, \quad r_0 \leq 2^*/(2^*)' \end{cases}$$

$$2^* = \begin{cases} \dfrac{2n}{n-2} & \text{if } n > 2 \\ q \in R & \text{if } n = 2 \end{cases} \qquad p' = \frac{p}{p-1}$$

while the functions f_1, f_0, g lie in suitable L^p-spaces, and that the ellipticity condition

$$A^{\alpha}_{ip_\beta j}(x,u,p)\xi^i_\alpha \xi^j_\beta \geq \nu|\xi|^2 \qquad \forall \xi ;\ \nu > 0$$

holds.

We note that minimum points or, more generally, stationary points of integral functional

$$J[u] = \int_\Omega F(x,u,\nabla u)\,dx$$

in the class of admissible functions

$$K = \{u \in H^1(\Omega, R^n) : u = u_0 \ \text{ on } \ \partial\Omega,\ \ \text{div } u = g \ \text{ in } \ \Omega\}$$

assumed nonempty, satisfy a system of the type (1.5) provided $F(x,u,p)$ fulfills suitable assumptions.

Moreover, due to the growth assumptions in (1.6), it is easily seen that the classical Navier-Stokes system

$$\text{div } u = 0$$

$$u \cdot \nabla u = \nu \nabla u - \text{grad } p + f$$

in its weak formulation is included in (1.5) provided $n \leq 4$.

We shall not present such results and we only refer the interested readers to [123].

Parabolic systems. The partial and everywhere regularity of solutions of nonlinear parabolic systems has been studied following the lines of the elliptic case.

Under controllable growth conditions, using the indirect approach of Chapter IV, partial regularity for solutions of second order quasilinear parabolic systems was proved by M. Giaquinta, E. Giusti [112] and extended to higher order systems by G. N. Daniljuk and I. V. Skrypnik [68].

The direct approach of Chapter VI, which permits to handle both controllable and natural growth conditions, has been carried on (the main point is the L^p-estimate) in M. Giaquinta-M. Struwe [128]; related everywhere regularity results for diagonal systems can be found in [289][290] [127].

For the existence and regularity of the parabolic flow associated with harmonic mappings we refer to [78][148][179][127].

We mention papers [54][55][56] where nonlinear parabolic systems under general controllable growth conditions are studied.

Finally we would like to quote the result of L. Caffarelli, R. Kohn, L. Nirenberg [41], which gives a relevant partial regularity result for solutions of the Navier-Stokes system

2. *Boundary regularity*

Roughly speaking, most of the regularity results we have stated in the interior hold also up to the boundary at least for the Dirichlet boundary value problem and provided $\partial\Omega$ be smooth, and for those we refer to the quoted papers.

Anyway we must say that the boundary regularity has not been studied very much, especially in respect to general boundary value problems.

Here we want to make only a few remarks on the possibility of extending the methods used for the interior regularity to proving regularity up to the boundary.

(a) First let us consider linear systems. The results in Chapter III can be straightforwardly extended up to the boundary, at least for the Dirichlet problem, see [45], and for the Neumann type problems, see for example [123].

We note anyway that this extension has not been carried on for 'general' boundary value problems for higher order systems.

(b) Step (a) allows us without strong difficulties to extend up to the boundary the partial regularity results stated in the interior. As an example of a theorem that can be obtained we state

THEOREM 2.1. *Let* u *be a weak solution to the Dirichlet problem*

$$-D_\beta[A^{\alpha\beta}_{ij}(x,u)D_\alpha u^i] = 0 \qquad j = 1,\cdots,N \quad \text{in } \Omega$$

$$u = \phi \qquad\qquad \text{on } \partial\Omega$$

with ϕ *smooth and* $A^{\alpha\beta}_{ij}$ *continuous functions satisfying*

$$|A^{\alpha\beta}_{ij}| \le L \qquad A^{\alpha\beta}_{ij}\xi^i_\alpha\xi^j_\beta \ge \nu|\xi|^2 \qquad \forall\xi;\ \nu > 0$$

Then u *is Hölder-continuous up to the boundary except for a closed singular set* Σ_0 *in* Ω *and a closed singular set* Σ_1, *on* $\partial\Omega$, *whose Hausdorff dimension does not exceed* $n-q$, *for some* $q > 2$. *Moreover, there exists* $\varepsilon_0 > 0$:

$$\Sigma_0 \cup \Sigma_1 \subset \{x : \sup_{B_R(x)\cap\Omega} |u| = +\infty\} \cup \left[x_0 \in \bar{\Omega} : \liminf_{R\to 0} \int_{B_R(x_0)\cap\Omega} |\nabla u|^2 dx > \varepsilon\right]$$

We mention that Theorem 2.1 is included in a more general theorem proved in [63] by using the indirect method of Chapter IV.

Theorem 2.1 immediately poses the two following questions:

1. Can the singular set Σ_1 be actually nonvoid?

2. If so, can the estimate of the Hausdorff measure of Σ_1, be improved? Note that the singular set Σ_1 on the (n–1)-manifold $\partial\Omega$ from Theorem 2.1 happens to be as 'large' as Σ_0 on the n-dimensional open set Ω.

In [105] an example is given which shows that Σ_1 can be nonvoid and that the Hausdorff estimate is 'optimal.'

(c) Some of the results in Chapter VII have been extended up to the boundary (Dirichlet problem), as we have mentioned, but there is no general treatment of the boundary regularity.

For the results in Section 3, Chapter VII, we refer to [165][119][306][212], and for the ones referring to harmonic mappings of manifolds to [163][119] and especially to [262][181].

Chapter IX
DIRECT METHODS FOR THE REGULARITY

In this chapter we shall describe some recent results on the existence and regularity of minima of nondifferentiable functionals

$$J[u,\Omega] = \int_{\Omega} F(x,u,Du)\,dx \,.$$

Except for the results in [114] (presented in Chapter III Section 3, Chapter VI Section 3 and Chapter VII Section 2) all the previous regularity results for minima of regular functionals have as a starting point the Euler equation of the functional in consideration.

As we have already remarked, this approach presents many inconveniences:

a) It requires some smoothness of F, moreover suitable growth conditions, not only on F, but also on its partial derivatives F_u and F_p.

b) Under natural growth conditions we need to start with bounded minimum points u, and also assume, in the vector valued case, some smallness condition on u. This often does not permit us to apply the results to minimum points which we are able to find in general only in $H^{1,2}$.

In conclusion this approach does not distinguish between true minima and simple extremals.

Starting with [114], M. Giaquinta and E. Giusti [116][117] have tried to develop a theory of regularity for minimum points, working directly with the functional J instead of working with its Euler equation.

This, together with some improvements of the existence theory, is the subject of this chapter.

1. *Quasi-minima*

Let us consider the functional

$$J[u,\Omega] = \int_\Omega F(x,u,Du)\,dx \tag{1.1}$$

where $F:\Omega\times R^N\times R^{nN}\to R$ is a Caratheodory function satisfying

$$|p|^m - b|u|^\gamma - g(x) \le F(x,u,p) \le \mu|p|^m + b|u|^\gamma + g(x) \tag{1.2}$$

with

$$1 \le m < n$$

$$1 \le \gamma < m^* = \frac{mn}{n-m}\,.$$

DEFINITION. *We say that* $u \in H^{1,m}_{\mathrm{loc}}(\Omega, R^N)$ *is a quasi-minimum* (*Q-minimum*) *for* J *in* Ω (*with constant* Q) *if*

$$J[u;\operatorname{supp}\phi] \le Q\,J[u+\phi;\operatorname{supp}\phi] \tag{1.3}$$

for all ϕ *with* $\operatorname{supp}\phi \subset\subset \Omega$.

Then, with small changes in the proof of Theorems 2.1, 2.2, Chapter VII and Theorem 3.1, Chapter V, we have (see [117])

THEOREM 1.1. *Let* u *be a* Q*-minimum for* J *in* Ω. *Then*

i) *if* $N=1$ *and* $g \in L^s(\Omega)$ *for some* $s > \frac{n}{m}$, *then* u *is locally Hölder-continuous, in particular locally bounded*

ii) *if* $N \ge 1$ *and* $g \in L^s(\Omega)$ *for some* $s>1$, *then there exists an exponent* $r>m$ *such that* $u \in H^{1,r}_{\mathrm{loc}}(\Omega, R^N)$.

Note that we are not assuming $F(x,u,p)$ convex in p, nor regularity of F and growth conditions on the derivatives of F.

Quite a lot of basic regularity results for solutions of (linear and nonlinear) elliptic systems can be now reread in terms of Q-minima.

1. Of course any minimum point u of J in (1.1) is a Q-minimum. Moreover it is not difficult to see that u is also a Q-minimum for

$$\int_\Omega \{|Du|^m + b|u|^\gamma + (b+g)\}\,dx\ .$$

In particular for $m=2$, $b=0$, $g=0$ it is a Q-minimum for the Dirichlet integral.

Any weak solution of the linear elliptic system with L^∞ coefficients $A^{\alpha\beta}_{ij}(x)$:

$$-D_\beta(A^{\alpha\beta}_{ij}(x)D_\alpha u^i) = 0 \quad j=1,\cdots,N\,, \qquad A^{\alpha\beta}_{ij}\xi^i_\alpha\cdot\xi^j_\beta \geq |\xi|^2 \quad \forall\xi\in R^{nN}$$

is a Q-minimum for the Dirichlet integral. To see that, it is sufficient to test with $u-v$, $\operatorname{supp}(u-v)\subset\subset\Omega$. In particular for $N=1$ we get De Giorgi's result, Theorem 2.1, Chapter II.

More generally, weak solutions of nonlinear elliptic systems under natural and usual hypotheses are Q-minima. Thus the Hölder-continuity of weak solutions to a large class of nonlinear elliptic equations (compare with [190]) and almost all the L^p-estimates for the gradient of general elliptic systems of Chapter V can be obtained as consequence of Theorem 1.1. In fact we have:

2. Let u be a weak solution of

$$(1.4) \qquad \int_\Omega [A^\alpha_i(x,u,Du)D_\alpha\phi^i + B_i(x,u,Du)\phi^i]\,dx = 0 \qquad \forall\phi\in C^\infty_0(\Omega,R^N)\ .$$

(A) Suppose that the *controllable growth conditions* and the ellipticity hold in the following weak form:

$$A_i^{\alpha}(x,u,p)\, p_\alpha^i \geq |p|^m - L|u|^\gamma - f(x) \qquad \gamma < m^*$$

$$|A(x,u,p)| \leq L|p|^{m-1} + L|u|^\sigma + g(x) \qquad \sigma = \gamma \frac{m-1}{m}$$

$$|B(x,u,p)| \leq L|p|^\tau + L|u|^\delta + h(x) \qquad \tau = \frac{\gamma-1}{\gamma}\, m\,, \quad \delta = \gamma - 1\,.$$

Then choosing $\phi = u - v$, with $\operatorname{supp}(u-v) \subset\subset \Omega$, we get that u *is a* Q-*minimum for*

$$\int_\Omega \left[|Du|^m + |u|^\gamma + \left(f + g^{\frac{m}{m-1}} + h^{\frac{\gamma}{\gamma-1}} + 1\right)\right] dx\,.$$

(B) Suppose that the *natural growth conditions* hold

$$\text{(1.5)} \qquad \begin{aligned} &A_i^{\alpha}(x,u,p)\, p_\alpha^i \geq |p|^m - L - Lf(x) \\ &|A(x,u,p)| \leq L|p|^{m-1} + L + Lg(x) \\ &|B(x,u,p)| \quad a|p|^m \quad + L + Lh(x) \\ &L = L(M)\,, \quad a = a(M)\,, \quad |u| \leq M\,. \end{aligned}$$

(B_1) *Suppose moreover that* $N = 1$. Then we get that *a (bounded) weak solution* u *is a* Q-*minimum for*

$$\text{(1.6)} \qquad \int_\Omega \left[|Du|^m + \left(f + g^{\frac{m}{m-1}} + h + 1\right)\right] dx\,.$$

This can be shown by choosing $\sup(u-w, 0)\, e^{\lambda(u-w)}$ and $\sup(w-u, 0)\, e^{\lambda(w-u)}$ as test function ϕ, where $w = v$ for $|v| \leq M$, $w = -M$ for $v \leq -M$, $w = M$ for $v \geq M$, for any v with $\operatorname{supp}(u-v) \subset\subset \Omega$.

(B_2) As we have seen, the L^p-estimate of $|Du|$ is not true under (1.5) if $a(M)M > 1$. But if *we assume* $2a(M)M < 1$, *then any weak solution* u, $|u| \leq M$, *of* (1.4) *is a* Q-*minimum for* (1.6). Therefore the L^p-estimate holds.

Let us mention two more examples:

3. Weak solution of the obstacle problem. Let u be a weak solution of the variational inequality

$$u \geq \psi \text{ in } \Omega : \int DuD(u-v)\,dx \leq 0 \quad \forall v,\ v \geq \psi,\ \operatorname{supp}(u-v) \subset\subset \Omega$$

then u is a Q-minimum for

$$\int_\Omega [|Du|^2 + |D\psi|^2]\,dx\ .$$

4. Let $u : \Omega \subset R^n \to R^n$ be a quasi-regular mapping, i.e. $u \in H^{1,n}_{loc}(\Omega, R^n)$ and for almost every $x \in \Omega$

$$|Du(x)|^n \leq k \det Du(x)\ .$$

Noting that for $\phi \in H^{1,n}_0(\Omega, R^n)$

$$\int_\Omega \det D\phi\,dx = 0$$

we get that u is a Q-minimum for

$$\int_\Omega |Du|^n dx\ .$$

Therefore we have $u \in H^{1,n+\varepsilon}_{loc}(\Omega, R^n)$, in particular u is locally Hölder-continuous (compare e.g. with [103]).

For more details on the proofs of the statements above, as well as for more information on the properties of Q-minima, we refer to [117]. Here we want to underline the unifying character of this notion: the first basic regularity results (Hölder-continuity in the scalar case $N = 1$, L^p-estimates for the gradient in the general case $N \geq 1$) are a consequence of the minimality condition (1.3) and not of the convexity (or ellipticity) of the functional. For systems, (1.3) is essentially a consequence of the ellipticity and of the growth conditions.

It is worth noting that, in the vector valued case, there is no hope to develop a Hölder regularity theory (even partial) for quasi-minima. Example 1.1 below, in fact, shows that there exists a quasi-minimum for the Dirichlet integral which is singular in a dense set.

Let $u = (u^1, \cdots, u^n)$ be a Q-minimum in Ω for the Dirichlet integral and suppose that $Q = 1 + \varepsilon$. Of course if ε is sufficiently small, then u is locally Holder-continuous in Ω. In analogy with the definition in [5], [32], G. Anzellotti [7] has considered quasi-minima in the following sense:

$$\int_{B_R} |Du|^2 dx \leq [1 + \omega(R)] \int_{B_R} |D(u + \phi)|^2 dx \qquad \forall \phi \in H^1(B_R, R^n), B_R \subset\subset \Omega .$$

He proves that the first derivatives of quasi-minima are locally Hölder-continuous, with exponent γ, in Ω provided $0 \leq \omega(R) \leq cR^{2\gamma}$.

EXAMPLE 1.1 ([274]). Let us start with a few remarks. Set

$$a_{ij}^{\alpha\beta}(x) = \delta_{ij}\delta^{\alpha\beta} + \frac{d_i^\alpha d_j^\beta}{\sum_{s,\gamma=1}^{n} u_{x_\gamma}^s \cdot d_s^\gamma} \qquad \alpha, \beta, i, j = 1, \cdots, n$$

where

$$d_i^\alpha = b_i^\alpha - u_{x_\alpha}^i , \quad b_i^\alpha \in L^2(\Omega) , \quad \int_\Omega b_i^\alpha D_\alpha \phi^i dx = 0 \qquad \forall \phi \in C_0^\infty(\Omega, R^n) .$$

Then

$$\int_{\Omega} a_{ij}^{\alpha\beta}(x)\, u^i_{x_\alpha} \phi^j_{x_\beta}\, dx = 0 . \tag{1.6}$$

The ellipticity and boundedness of the coefficients $a_{ij}^{\alpha\beta}$ corresponds respectively to

$$u_x \cdot d > 0$$

$$\frac{b \cdot d}{u_x \cdot d} \leq M .$$

It is on the basis of this simple remark that the examples 3.1, 3.2 of Chapter II can be regarded. Actually, the following choice for $n \geq 3$

$$\begin{aligned} u(x) &= |x|^{-1} x \\ b_i^\alpha(x) &= |x|^{-1}\left(n\,\delta_{i\alpha} + \frac{n}{n-2}\frac{x_i x_\alpha}{|x|^2}\right) \end{aligned} \tag{1.7}$$

permits to construct a discontinuous weak solution of the elliptic system (1.6).

Let y_h be a sequence of points in Ω and let us set

$$U^i(x) = \sum_h u^i(x-y_h)\, \varepsilon_h$$

$$B_i^\alpha(x) = \sum_h b_i^\alpha(x-y_h)\, \varepsilon_h$$

and

$$A_{ij}^{\alpha\beta}(x) = \delta_{ij}\delta^{\alpha\beta} + \frac{D_i^\alpha D_j^\beta}{U_x \cdot D}$$

where

$$D_i^\alpha = B_i^\alpha - U^i_{x_\alpha}$$

and u, b are defined by (1.7).

Since

$$\int B_i^\alpha \phi^i_{x_\alpha} dx = 0 \qquad \forall \phi \in C_0^\infty(\Omega, R^n)$$

it is a simple matter of calculation to show that after a suitable choice of the ε_h, the vector U belongs to $H^{1,2}_{loc}(\Omega, R^n)$ and is a solution of the elliptic system

$$-D_\beta(A_{ij}^{\alpha\beta} D_\alpha U^i) = 0 \qquad j = 1, \cdots, n .$$

Choosing the sequence y_h we may have, of course, U singular in a dense set.

2. *Quasi-minima and quasi-convexity*

At the end of Section 2, Chapter I, we remarked that in the semicontinuity theorem 2.3 the convexity assumption is natural in the scalar case $N = 1$; actually it is necessary; but it is very far from being necessary in the vector valued case $N > 1$. It should be substituted with the *quasi-convexity condition* of C. B. Morrey [231, Sec. 4.4].

The Carathéodory function $F: \Omega \times R^N \times R^{nN} \to R$ *is called quasi-convex if for* a.e. $x_0 \in \Omega$ *and for all* $u_0 \in R^N$, $\xi_0 \in R^{nN}$ *we have*

$$\frac{1}{|\Omega|} \int_\Omega F(x_0, u_0, \xi_0 + D\phi(x))\, dx \geq F(x_0, u_0, \xi_0) \qquad \forall \phi \in C_0^\infty(\Omega, R^N)$$

i.e. if the frozen functional

$$J^0[u; \Omega] = \int_\Omega F(x_0, u_0, Du(x))\, dx$$

has the linear functions as minimum points.

Quasi-convexity is weaker than convexity and it reduces to it for $n \geq 2$ $N = 1$ or $n = 1$ $N \geq 1$. Examples of quasi-convex functions are given by convex functions of the invariants of the Jacobian matrix of u, see [231]; compare also with [19].

Semicontinuity theorems under the quasi-convexity condition plus quite strong assumptions have been proved in [224][215][231] and in [19][23]. Recently N. Fusco and E. Acerbi-N. Fusco have given almost optimal semicontinuity theorems in [101][1].

Let us state the main theorem of [1] without proof.

THEOREM 2.1. *Let* $F(x,u,p)$ *be measurable in* $x \in \Omega$ *for all* (u,p) *and continuous in* (u,p) *for* a.e. $x \in \Omega$. *Assume that*

$$0 \le F(x,u,p) \le 1 + \lambda(|u|^m + |p|^m) \qquad m \ge 1\,. \tag{2.1}$$

Then the functional

$$\int_\Omega F(x,u,Du)\,dx \tag{2.2}$$

is weakly s.l.s.c. *in* $H^{1,m}(\Omega, R^N)$ *if and only if* F *is quasi-convex.*

The proof is not very simple; instead, it is easier to prove weak semicontinuity of (2.2) in any space $H^{1,q}(\Omega, R^N)$ with $q > m$, see [101][207], even under the weaker assumption

$$|F(x,u,p)| \le 1 + \lambda(|u|^m + |p|^m) \qquad m \ge 1\,. \tag{2.3}$$

More precisely we have (see [101][207] for the proof):

THEOREM 2.2. *Let* $F(x,u,p)$ *be measurable in* x *and continuous in* (u,p). *Assume that (2.3) holds and that* F *be quasi-convex. Then the functional (2.2) is weakly* s.l.s.c. *in* $H^{1,q}(\Omega, R^N)$ *for any* $q > m$.

We note that Theorem 2.2 fails if $q = m$, as an example of F. Murat and J. L. Tartar [236] shows.

Now let us consider for the sake of simplicity the functional

$$J[u;\Omega] = \int_\Omega F(x,u,Du)\,dx \tag{2.4}$$

under the growth condition

$$|p|^m \leq F(x,u,p) \leq \mu|p|^m \qquad m > 1 \tag{2.5}$$

and let us recall the following variational principle in Ekeland [81]:

THEOREM 2.3. *Let* (V,d) *be a complete metric space,* $J : V \to [0,+\infty]$ *a lower semicontinuous functional,* $J \not\equiv +\infty$. *Let* $\eta > 0$ *and* $w \in V$ *satisfy*

$$J(w) \leq \inf_V J + \eta .$$

Then there exists $v \in V$ *such that* $J(v) \leq J(w)$, $d(v,w) \leq 1$ *and* v *is the only minimum point for the functional*

$$G(u) = J(u) + \eta\, d(u,v) .$$

The functional (2.4), under (2.5), is obviously semicontinuous (actually continuous) in the metric space $\{u \in H^{1,1}(\Omega, R^N) : u = \tilde{u} \text{ on } \partial\Omega\}$ for a given $\tilde{u}$ (for example in $H^{1,m}(\Omega, R^N)$). Hence we may apply Theorem 2.3, and the function v we obtain is clearly a Q-minimum for the functional

$$\int_\Omega (1 + |Dz|^m)\,dx$$

with constant Q independent of η for η small. In particular we may conclude:

There exists a minimizing sequence $\{u_h\}$ *for* $J[u;\Omega]$ *in* $\{u \in H^{1,1}(\Omega, R^N) : u = \tilde{u}$ *on* $\partial\Omega\}$ *made of* Q*-minima with constant* Q *uniform.*

Theorem 1.1 then implies that we can bound the $H^{1,r}$ norm (for some $r > m$) of the u_h's on $\tilde{\Omega} \subset\subset \Omega$ with a constant depending on $\tilde{\Omega}$ but not on h.

If we now assume that F in (2.4) is quasi-convex, by means of Theorem 2.2, we conclude:

THEOREM 2.4. *Let* $\tilde{u} \in H^{1,m}(\Omega, R^N)$ *and let* F *be a quasi-convex function satisfying (2.5). Then there exists a minimum point for the functional (2.4) in* $H^{1,m} \cap \{u : u = \tilde{u} \text{ on } \partial\Omega\}$. *Moreover* $u \in H^{1,r}_{loc}(\Omega, R^N)$ *for some* $r > m$.

The proof of Theorem 2.4 we have given above is a rereading of the proof in [207].

We notice that there is no known regularity result (in the sense of Hölder or partial Hölder continuity) for the minimum points in Theorem 2.4 (except obviously for the case $m \geq n$).

3. *The singular set of minima of a class of quadratic functionals*

Let us consider the quadratic functional

$$J[u;\Omega] = \int_\Omega A^{\alpha\beta}_{ij}(x,u) D_\alpha u^i D_\beta u^j \, dx \qquad (A^{\alpha\beta}_{ij} = A^{\beta\alpha}_{ji}) \tag{3.1}$$

where the coefficients $A^{\alpha\beta}_{ij}$ are

(i) bounded: $|A(x,z)| \leq M$

(ii) elliptic: $A\xi\cdot\xi = A^{\alpha\beta}_{ij}(x,z)\xi^i_\alpha \xi^j_\beta \geq |\xi|^2 \quad \forall \xi$

(iii) (uniformly) continuous: $|A(x,z) - A(x',z')| \leq \omega(|x-x'|^2 + |z-z'|^2)$

$\omega(t)$ being a bounded continuous concave function with $\omega(0) = 0$.

In Section 3 of Chapter VI we have seen that minimum points of $J[u;\Omega]$ are Hölder-continuous in an open set Ω_0 and that the Hausdorff dimension of the singular set is strictly less than $n-2$. Moreover we have seen that the minimum points are as regular in Ω_0 as the regularity of the A's permits.

In this section we want to show that for a special class of quadratic multiple integrals and bounded minimum points we can improve the estimate of the Hausdorff dimension of the singular set.

More precisely we shall restrict ourselves to the special form of the coefficients given by

$$\text{(iv)} \qquad A_{ij}^{\alpha\beta}(x,u) = g_{ij}(x,u)\, G^{\alpha\beta}(x)$$

moreover we shall assume that the function ω in (iii) satisfies:

$$\text{(v)} \qquad \int_0^1 \frac{\omega(t^2)}{t}\, dt < +\infty .$$

Then we have, see [115]

THEOREM 3.1. *Let (i) (ii) (iii) (iv) (v) hold and let* u *be a bounded minimum of the functional* J. *Then*

a. *if* $n = 3$, u *may have at most isolated singular points*

b. *if* $n \geq 4$, *the dimension of the singular set of* u *cannot exceed* $n-3$.

The proof uses some ideas taken from the regularity theory for parametric minimal surfaces plus the result and some estimates of Section 3, Chapter VI.

Let us state the main points of the proof.

The first lemma is a result concerning the convergence of functionals and minima. It could be stated for general functionals of the type (1.1).

LEMMA 3.1. *Let* $A^{(\nu)}(x,z) = A_{ij}^{\alpha\beta(\nu)}(x,z)$ *be a sequence of continuous functions in* $B_1(0) \times R^N$ *converging uniformly to* $A(x,z)$ *and satisfying (i) (ii) (iii) (uniformly with respect to* ν *). For each* $\nu = 1, 2, \cdots$ *let* $u^{(\nu)}$ *be a minimum point in* $B_1(0)$ *for*

$$J^{(\nu)}[u^{(\nu)}; B] = \int_B A^{(\nu)}(x,u) Du^{(\nu)} Du^{(\nu)} dx \,, \qquad B = B_1(0)$$

suppose that $u^{(\nu)}$ *converges weakly in* $L^2(B, R^N)$ *to* v. *Then* v *is a minimum for*

$$J[u,B] = \int_B A(x,u) Du\, Du\, dx \,.$$

Moreover, if x_ν *is a singular point for* $u^{(\nu)}$, *and* $x_\nu \to x_0$, *then* x_0 *is a singular point for* v.

Proof. We know, see Section 3, Chapter V, that for each ball $B_r = B_r(x_0) \subset B$ we have

$$\int_{B_{r/2}} |Du^{(\nu)}|^2 dx \le \gamma_1 r^{-2} \int_{B_r} |u^{(\nu)} - u_r^{(\nu)}|^2 dx \tag{3.2}$$

where

$$u_r^{(\nu)} = \fint_{B_r} u^{(\nu)} dx$$

and that there exists a $q > 2$, independent of ν, such that

$$\left(\fint_{B_{r/2}} |Du^{(\nu)}|^q dx \right)^{1/q} \le \gamma_2 \left(\fint_{B_r} |Du^{(\nu)}|^2 dx \right)^{1/2} . \tag{3.3}$$

It follows that $Du^{(\nu)} \in L^q_{loc}(B)$ and that for every $R < 1$

$$\int_{B_r} |Du^{(\nu)}|^q dx \le c(R) \,. \tag{3.4}$$

This together with the weak L^2-convergence implies, passing possibly to a subsequence, that

$$J[v;B_R] \leq \liminf_{\nu \quad \infty} J^{(\nu)}[u^{(\nu)};B_R]\,. \tag{3.5}$$

Let now w be an arbitrary function coinciding with v outside B_R, and let $\eta(x) \in C^1(B)$, $0 \leq \eta \leq 1$, $\eta = 0$ in $B_\rho(0)$, $\rho \leq 1$, and $\eta \equiv 1$ outside B_R. Then $v^{(\nu)} \equiv w + \eta(u^\nu - v)$ coincide with $u^{(\nu)}$ outside B_R, therefore

$$J^{(\nu)}[u^{(\nu)};B_R] \leq J^{(\nu)}[v^{(\nu)};B_R]\,. \tag{3.6}$$

Taking (i) and (3.3) into account we get

$$J^{(\nu)}[v^{(\nu)};B_R] \leq \int_{B_R} A^{(\nu)}(x,v^{(\nu)})\,Dw\,Dw + \gamma_3(R)\,\|\eta\|_{\frac{q}{q-2},B_R} +$$

$$+ \gamma_4(R,\eta)\,\|u^{(\nu)} - v\|_{2,B_R}(1 + \|u^{(\nu)} - v\|_{2,B_R})$$

and letting $\nu \to \infty$, we deduce from (3.5)(3.6)

$$J[v;B_R] \leq J[w;B_R] + \gamma_3\|\eta\|_{\frac{q}{q-2},B_R}\,.$$

Taking ρ close to R, the last term can be made arbitrarily small, and that proves the first assertion of the lemma.

In order to prove the second part of the lemma, let us recall that, because of Caccioppoli's inequality (3.2), a point $\bar{x}$ is singular if and only if

$$\liminf_{\rho \to 0} \rho^{-n} \int_{B_\rho(\bar{x})} |u - u_{\bar{x},\rho}|^2 \geq \varepsilon_0^2$$

where ε_0 depends only on ω and therefore is independent of ν (compare with Theorem 3.1 and 1.1 of Chapter VI).

Suppose now that x_0 is a regular point for v. Then for ρ small enough we have

$$\rho^{-n} \int_{B_\rho(x_0)} |v-v_\rho|^2 dx < \varepsilon_0^2$$

and hence

$$\lim_{\rho\to\infty} \rho^{-n} \int_{B_\rho(x^{(\nu)})} |u^{(\nu)} - u_\rho^{(\nu)}|^2 dx = \rho^{-n} \int_{B_\rho(x_0)} |v-v_\rho|^2 < \varepsilon_0^2$$

which implies that $x^{(\nu)}$ is a regular point for $u^{(\nu)}$, provided ν is large enough. This concludes the proof of the lemma. q.e.d.

The second lemma is a monotonicity result like the well-known one for minimal surfaces. The special structure of the coefficients (iv) (v) is needed only to prove this lemma. Any extension of the lemma to a more general class of coefficients will therefore permit an extension of Theorem 3.1.

We may (and do) assume without loss in generality that

$$G^{\alpha\beta}(0) = \delta_{\alpha\beta} .$$

Then the monotonicity lemma is:

LEMMA 3.2. *Let (i) ... (v) hold and let* u *be a minimum for* $J[v; B_1(0)]$. *Then for every* ρ, R, $0 < \rho < R < 1$, *we have*

$$\int_{\partial B_1} |u(Rx) - u(\rho x)|^2 d\mathcal{H}^{n-1}(x) \le \gamma_5 \log\left(\frac{R}{\rho}\right) [\Phi(R) - \Phi(\rho)] \tag{3.7}$$

where

$$\Phi(t) = t^{2-n} \exp\left(\gamma_6 \int_0^t \frac{\omega(s^2)}{s}\, ds\right) \int_{B_t} A(x,u)\, Du\, Du\, dx\ .$$

Proof. For the sake of simplicity let us assume moreover that the coefficients do not depend explicitly on x, i.e.

$$A_{ij}^{\alpha\beta}(x,z) = \delta_{\alpha\beta}\, g_{ij}(z)\ .$$

In this case $\Phi(t)$ reduces to

$$\Phi(t) = t^{2-n} \int_{B_t} A(u)\, Du\, Du\, dx\ .$$

We refer to [115] for the general case.

For $t<1$ let $x_t = t\dfrac{x}{|x|}$ and $u_t(x) = u(x_t)$. We have

$$J[u; B_t] \le J[u_t; B_t] \tag{3.8}$$

and

$$J[u_t; B_t] =$$

$$= \int_{B_t} A_{ij}^{\alpha\beta}(u(x_t))\, \frac{t^2}{|x|^2} \left(\delta_{\alpha h} - \frac{x_\alpha x_h}{|x|^2}\right)\left(\delta_{\beta k} - \frac{x_\beta x_k}{|x|^2}\right) D_h u^i(x_t)\, D_k u^j(x_t)\, dt\ .$$

Observing that for every f we have

$$\int_{B_t} |x|^{-2} f(x_t)\, dx = \frac{t^{-1}}{n-2} \int_{\partial B_t} f(x)\, d\mathcal{H}^{n-1}$$

we get

$$J[u_t; B_t] =$$

$$= \frac{t}{n-2}\left\{\int_{\partial B_t} A(u)DuDu\, d\mathcal{H}^{n-1} - \int_{\partial B_t} A^{\alpha\beta}_{ij}\frac{x_\alpha x_h}{|x|^2}\left[2\delta_{\beta k} - \frac{x_\beta x_k}{|x|^2}\right] D_h u^i D_k u^j d\mathcal{H}^{n-1}\right. .$$

Taking into account the special form of the coefficient and the ellipticity, we conclude that

$$J[u_t; B_t] \le$$

$$\le \frac{t}{n-2}\left\{\int_{\partial B_t} A(u)DuDu\, d\mathcal{H}^{n-1} - \int_{\partial B_t} \frac{|<x, Du>|^2}{|x|^2}\, d\mathcal{H}^{n-1}\right\} \tag{3.9}$$

$$<x, Du> = x_\alpha D_\alpha u .$$

Now we have

$$t^{2-n}\int_{\partial B_t} A(u)DuDu\, d\mathcal{H}^{n-1} = \phi'(t) + (n-2)\frac{\phi(t)}{t}$$

therefore from (3.9), (3.8) we get

$$\phi'(t) \ge t^{2-n}\int_{\partial B_t} \frac{|<x, Du>|^2}{|x|^2}\, d\mathcal{H}^{n-1}$$

and integrating

$$\phi(R) - \phi(\rho) \ge \int_\rho^R t^{2-n}\int_{\partial B_t} \frac{|<x, Du>|^2}{|x|^2}\, d\mathcal{H}^{n-1} .$$

On the other hand

$$|u(Rx) - u(\rho x)|^2 \le \left(\int_\rho^R |<x, Du(tx)>| \, dt\right)^2 \le$$

$$\log\left(\frac{R}{\rho}\right) \int_\rho^R t|<x, Du(tx)>|^2 dt$$

from which the conclusion follows at once integrating on ∂B. q.e.d.

Proof of Theorem 3.1. a). Suppose that u has a sequence of singular points, x_ν, converging to $x_0 = 0$. We use a rescaling argument. Let $R_\nu = 2|x_\nu| < 1$. The function $u^{(\nu)} = u(R_\nu x)$ is a local minimum point in B for

$$J^{(\nu)}[u^\nu; B] = \int_B A^{(\nu)}(x, u^{(\nu)}) Du^{(\nu)} Du^{(\nu)} dx$$

$$A^\nu(x,z) = A(R_\nu x, z) .$$

Moreover, each $u^{(\nu)}$ has a singular point y_ν with $|y_\nu| = \frac{1}{2}$. Since the $u^{(\nu)}$ are uniformly bounded, we can suppose that they converge weakly in $L^2(B)$ to some function v and $y_\nu \to y_0$. Now we may apply Lemma 3.1 and conclude that v is a local minimum point for

$$J^0[v; B] = \int_B A(0,v) Dv\, Dv\, dy .$$

Also from Lemma 3.1 it follows that v has a singular point at y_0. Now we claim that v *is homogeneous of degree zero.* This is a consequence of the monotonicity lemma. In fact first of all, from (3.7) it follows that $\phi(t)$ is increasing and therefore tends to a finite limit when $t \to 0$; secondly for $\rho = \lambda R_\nu$, $R = \mu R_\nu$ $0 < \lambda < \mu < 1$ we have

$$\int_{\partial B} |u^{(\nu)}(\lambda x) - u^{(\nu)}(\mu x)|^2 \, d\mathcal{H}^{n-1} \leq \gamma_5 \log\left(\frac{\mu}{\lambda}\right) [\phi(\mu R_\nu) - \phi(\lambda R_\nu)]$$

therefore, letting $\nu \to 0$, we conclude that

$$\int_{\partial B} |v(\lambda x) - v(\mu x)|^2 \, dx = 0$$

for almost every value λ and μ.
Since v is homogeneous of degree zero, the whole segment joining 0 with y_0 is made of singular points for v. This contradicts Theorem 3.1 Chapter VI, and in particular the conclusion that the singular set has Hausdorff dimension strictly less than $n-2 = 3-1 = 1$. q.e.d.

The proof of b) follows very closely the reduction argument used for minimal surfaces, compare with [86]. Here we do not give the proof and we refer to [115].

REMARK 3.1. Let us consider a mapping U from a Riemannian manifold X with metric tensor $G_{\alpha\beta}(x)$ into the Riemannian manifold M with metric tensor $g_{ij}(u)$. Then its 'energy' is given in local coordinates by:

$$E(u;A) = \int_A g_{ij}(u)\, G^{\alpha\beta}(x) D_\alpha u^i D_\beta u^j \sqrt{G}\, dx$$

where $(G^{\alpha\beta}) = (G_{\alpha\beta})^{-1}$, $G = \det(G_{\alpha\beta})$.

Theorem 3.1 gives information about the regularity of harmonic mapping (energy minimizing), i.e. minimum points of the functional $E(u;A)$ in case $X = \mathbf{R}^n$ with metric $G_{\alpha\beta}$ and $M = \mathbf{R}^N$ with metric g_{ij} without assumptions involving the sectional curvature of M. We mention that in this setting Theorem 3.1 has been proved (for the energy functional) for general Riemannian manifolds X, M by R. Schoen and K. Uhlenbeck [261]. For related results and more information we refer to [261][262][181][75].

4. $C^{1,\alpha}$*-regularity of minima*

In this section we investigate the Hölder and partial Hölder continuity of minimum points of nondifferentiable functionals

$$(4.1) \qquad J[u;\Omega] = \int_{\Omega} F(x,u,Du)\,dx$$

following [116].

We shall assume, for the sake of simplicity, that $F(x,u,p)$ *satisfies*

$$(4.2) \qquad \lambda|p|^2 \leq F(x,u,p) \leq \Lambda|p|^2 \qquad \lambda > 0$$

and moreover

i) *for every* $(x,u) \in \Omega\times R^N, F(x,u,p)$ *is twice differentiable in* p, *and we have*

$$(4.3) \qquad |F_{pp}(x,u,p)| \leq L$$

$$(4.4) \qquad F_{p^i_\alpha p^j_\beta}(x,u,p)\xi^\alpha_i\xi^\beta_j \geq |\xi|^2 \qquad \forall\xi \in R^{nN}.$$

ii) *For every* $p \in R^{nN}$ *the function* $(1+|p|^2)^{-1}F(x,u,p)$ *is continuous in* $\Omega\times R^N$ *uniformly in* p, *i.e. there exists a bounded nonnegative concave increasing function* $\omega(t)$, *with* $\omega(0)=0$, *such that*

$$(4.5) \qquad |F(x,u,p)-F(y,v,p)| \leq (1+|p|^2)\,\omega(|x-y|^2+|u-v|^2).$$

Note that we do not assume the existence (or any growth condition) of the derivatives F_u, and therefore our functionals are in general nondifferentiable.

The results will take a different form in the scalar and in the vector case. The technique, however, consists in both cases in comparing the minimum point u in a ball $B_R(x_0)$ with the function v minimizing the functional

$$J^0[v;B_R] = \int_{B_R(x_0)} F(x_0,u_{x_0,R},Dv)\,dx$$

among all functions in $H^{1,2}(B_R,R^N)$ taking the value u on ∂B_R.

The following lemma shall prove very useful.

LEMMA 4.1. *Let* v *minimize* J^0 *in* B_R *and let* $u=v$ *on* ∂B_R. *Then*

$$\int_{B_R} |Dw|^2\,dx \le 2\{J^0[u;B_R]-J^0[v;B_R]\} \tag{4.6}$$

$$w = u-v\ .$$

Proof. Set

$$F^0(p) = F(x_0, u_{x_0,R}, p)$$

we have, taking into account (4.4),

$$F^0(Du)-F^0(Dv) = F^0_{p^i_\alpha}(Dv)D_\alpha w^i + \int_0^1 (1-t)F^0_{p^i_\alpha p^j_\beta}(tDu+(1-t)Dv)D_\alpha w^i D_\beta w^j dt \ge$$

$$\ge F^0_{p^i_\alpha}(Dv)D_\alpha w^i + \frac{1}{2}|Dw|^2\ .$$

Integrating on B_R we immediately get (4.6) since v satisfies the Euler equation

$$\int_{B_R} F^0_{p^i_\alpha}(Dv)D_\alpha \phi^i\,dx = 0 \qquad \forall \phi \in H^1_0(B_R, R^N)\ . \tag{4.7}$$

q.e.d.

The scalar case. Let us start with the scalar case $N=1$. From the results of Section 2, Chapter VII, we know that every local minimum of $J[u;\Omega]$ is Hölder-continuous with some exponent $\gamma > 0$.

Now we show, in a similar way as in Section 1, Chapter VII, that under the above assumptions every local minimum of $J[u;\Omega]$ is Hölder-continuous with any exponent $\gamma < 1$. More precisely we have

THEOREM 4.1. *Let* u *be a minimum for the functional* J. *Then* Du *belongs to* $L^{2,n-\varepsilon}_{loc}(\Omega)$ *for every* $\varepsilon > 0$.

In particular, compare with Section 1, Chapter III, u belongs to $C^{0,\gamma}_{loc}(\Omega)$ for any $\gamma < 1$.

Proof. Let $B_R \subset\subset \Omega$ and let v minimize J^0 with boundary datum u on $\partial\Omega$. We have

$$\operatorname*{osc}_{B_R} v \le \operatorname*{osc}_{B_R} u \le c_1 R^{\alpha} \quad \text{for some} \quad \alpha > 0 . \tag{4.8}$$

On the other hand we have that $v \in H^{2,2}_{loc}(B_R)$ and differentiating (4.7):

$$\int_{B_R} F^0_{p_\alpha p_\beta}(Dv) D_\beta(D_s v) D_\alpha \phi \, dx = 0 \qquad \forall \phi \in H^1_0(B_R) .$$

Choosing $\phi = \eta D_s v$, $\eta \in C^\infty_0(B_R)$, and summing over s we conclude that $z = |Dv|^2$ is a subsolution of an elliptic operator:

$$\int_{B_R} F^0_{p_\alpha p_\beta}(Dv) D_\beta z D_\alpha \eta \le 0 \qquad \forall \eta \in C_0(B_R), \ \eta \ge 0 .$$

From the standard elliptic estimate (see e.g. [129]) we then get

$$\sup_{B_{R/2}} |Dv|^2 \le c_2 R^{-n} \int_{B_R} |Dv|^2 dx . \tag{4.9}$$

From (4.9) we obtain for every $\rho \le R/2$ the inequality

$$\int_{B_\rho} |Dv|^2 dx \le c_3 \left(\frac{\rho}{R}\right)^n \int_{B_R} |Dv|^2 dx$$

which we already had in Chapter VI.

Changing possibly the constant c_3, the above inequality holds for every $\rho < R$. Coming back to u we get

$$(4.10)\qquad \int_{B_\rho} |Du|^2 dx \le c_4 \left\{ \left(\frac{\rho}{R}\right)^n \int_{B_R} |Du|^2 dx + \int_{B_R} |D(u-v)|^2 dx \right\} .$$

In order to estimate the last integral we use (4.6). We have

$$\begin{aligned} J^0[u; B_R] - J^0[v; B_R] = & \int_{B_R} [F(x_0, u_{x_0,R}, Du) - F(x,u,Du)]\, dx + \\ & + \int_{B_R} [F(x,v,Dv) - F(x_0, u_{x_0,R}, Dv)]\, dx \\ & + J[u; B_R] - J[v; B_R] . \end{aligned}$$

Since u minimizes J in B_R, taking into account (4.5), (4.8) and the Hölder continuity of u we get

$$\int_{B_R} |D(u-v)|^2 dx \le c_5\, \omega_1(R) \int_{B_R} |Du|^2 dx$$

with $\omega_1(R) \downarrow 0$ as $R \downarrow 0$. In conclusion

$$\int_{B_\rho} |Du|^2 dx \le c_6 \left[\left(\frac{\rho}{R}\right)^n + \omega_1(R) \right] \int_{B_R} |Du|^2 dx$$

which implies by means of Lemma 2.1, Chapter III

$$(4.11)\qquad \int_{B_\rho} |Du|^2\,dx \le c_7 \left(\frac{\rho}{R_0}\right)^{n-\varepsilon} \int_{B_{R_0}} |Du|^2\,dx \qquad \rho < R_0 .$$

The conclusion then follows at once. q.e.d.

The next theorem deals with the Hölder-continuity of the first derivatives of minimum points u. For that, we have to make a further assumption on F, namely that F is Hölder-continuous in (x,u) with exponent 2σ. More precisely we shall assume that the function ω in (4.5) satisfies

$$(4.12)\qquad \omega(t) \le At^\sigma$$

for some $\sigma > 0$.

THEOREM 4.2. *Let (4.2), (4.3), (4.5), (4.12) hold. Then the first derivatives of* u *are Hölder-continuous.*

Proof. Let $B_R(x_0) \subset\subset \Omega$ and let v be as before. From De Giorgi's theorem (Theorem 2.1, Chapter II), we deduce (compare with Section 1, Chapter VII) the estimate

$$\int_{B_\rho(x_0)} |Dv - (Dv)_{x_0,\rho}|^2\,dx \le c_8 \left(\frac{\rho}{R}\right)^{n+2\delta} \int_{B_R(x_0)} |Dv - (Dv)_{x_0,R}|^2\,dx$$

for some $\delta > 0$. Hence we have

$$\int_{B_\rho(x_0)} |Du - (Du)_{x_0,\rho}|^2\,dx \le c_9 \left[\left(\frac{\rho}{R}\right)^{n+2\delta} \int_{B_R(x_0)} |Du - (Du)_{x_0,R}|^2 + \right.$$

$$\left. + \int_{B_R(x_0)} |D(u-v)|^2\,dx\right] .$$

The last integral can be estimated as above. Taking into account that $u \in C^{0,\alpha}$ we get

$$\int_{B_R} |D(u-v)|^2 dx \leq c_{10} R^{2\alpha\sigma} \int_{B_R} |Du|^2 dx$$

and using (4.11)

$$\int_{B_R} |D(u-v)|^2 dx \leq c_{11} R^{n+2\alpha\sigma-\varepsilon} .$$

Taking $\varepsilon = \alpha\sigma$ we get in conclusion

$$\int_{B_\rho(x_0)} |Du-(Du)_{x_0,\rho}|^2 dx \leq c_{12}\left(\frac{\rho}{R}\right)^{n+2\delta} \int_{B_R(x_0)} |Du-(Du)_{x_0,R}|^2 dx + c_{12} R^{n+\alpha\sigma}$$

for every $\rho < R$.

Using again Lemma 2.1, Chapter III we finally get

$$\int_{B_\rho} |Du-(Du)_{x_0,\rho}|^2 dx \leq c_{13}\rho^{n+2\tau}$$

$\tau = \min\left(\frac{\alpha\sigma}{2}, \frac{\delta}{2}\right)$ and hence the result (compare with Section 1, Chapter III).

q.e.d.

The vector valued case: partial regularity. In the general case $N > 1$ we expect only regularity in an open set Ω_0. Actually we have

THEOREM 4.3. *Let* u *be a minimum point for the functional (1.1). Suppose that (4.2), (4.3), (4.4) and (4.5) with* $\omega(t) \leq At^\sigma$ *for some* $\sigma > 0$ *hold. Then there exists an open set* $\Omega_0 \subset \Omega$ *such that* u *has Hölder-continuous first derivatives in* Ω_0. *Moreover we have* $\Omega - \Omega_0 = \Sigma_1 \cup \Sigma_2$, *where*

$$\Sigma_1 = \{x_0 \in \Omega : \sup_{r>0} |(Du)_{x_0,r}| = +\infty\}$$

$$\Sigma_2 = \left\{x_0 \in \Omega : \liminf_{r\to 0} r^{-n} \int_{B_r} |Du - (Du)_{x_0,r}|\, dx > 0\right\}$$

and hence $\operatorname{meas}(\Omega - \Omega_0) = 0$.

Theorem 4.3 has been proved in [116], and in a slight different way (more in the spirit of Section 2, Chapter VI) in [174]. The proof which follows is taken from [116].

Proof. We want to estimate

$$\Phi(x_0, \rho) = \int_{B(x_0,\rho)} |Du - (Du)_{x_0,\rho}|^2 dx .$$

Let $x_0 \in \Omega$, $R < \operatorname{dist}(x_0, \partial\Omega)$ and let v be the minimum point for $J^0[v; B_R]$ with boundary values $u(x)$ on ∂B_R.
Writing the equation in variation for Dv and using the regularity theory in Chapter VI we have: *for every* $L_0 > 0$ *there exists* $\eta_0(L_0) > 0$ (depending only on L_0, n, N and on the modulus of continuity of F_{pp}) *such that if*

$$(4.13) \qquad \fint_{B_R(x_0)} |Du|^2 dx \leq L_0^2 , \qquad \fint_{B_R(x_0)} |Dv - (Dv)_{x_0,R}|^2 dx < \eta_0^2$$

then Dv *is Hölder-continuous* (with every exponent $a < 1$) *in a neighborhood of* x_0, *and moreover*

$$(4.14) \qquad \int_{B_\rho(x_0)} |Dv-(Dv)_{x_0,\rho}|^2 dx \le c_{14}\Big(\frac{\rho}{R}\Big)^{n+2\alpha} \int_{B_R(x_0)} |Dv-(Dv)_{x_0,R}|^2 dx$$

for every $\rho < R$.

Now we want to show that (4.13) can be replaced by similar inequalities involving the function u. In fact, from (4.2) we have

$$(4.15) \qquad \fint_{B_R(x_0)} |Dv|^2 dx \le c_{15} \fint_{B_R(x_0)} |Du|^2 dx\,.$$

On the other hand from (4.7) rewritten as

$$\int_{B_R} [F_{p^i_\alpha}(Dv) - F_{p^i_\alpha}(\xi)] D_\alpha \phi^i dx = 0 \qquad \forall \phi \in H^1_0(B_R, R^N)$$

taking $\phi^i = v^i - u^i = v^i - \xi^i_\alpha x^\alpha - (u^i - \xi^i_\alpha x^\alpha)$, $(\xi^i_\alpha = (D_\alpha u^i)_{x_0,R})$ we get

$$(4.16) \qquad \fint_{B_R} |Dv-(Dv)_{x_0,R}|^2 dx \le c_{16} \fint_{B_R} |Du-(Du)_{x_0,R}|^2 dx\,.$$

From (4.15), (4.16) we conclude that *for every* $M_0 > 0$ *there exists* $\varepsilon_0(M_0)$ *such that the inequalities*

$$(4.17) \qquad \fint_{B_R(x_0)} |Du|^2 dx \le M_0^2 \qquad \Phi(x_0, R) < \varepsilon_0^2$$

imply (4.13) and therefore (4.14).

Suppose now that (4.17) are satisfied for some R. Then we have as in Theorem 4.2

$$\int_{B_\rho(x_0)} |Du-(Du)_{x_0,\rho}|^2\,dx \le c_{17}\left\{\left(\frac{\rho}{R}\right)^{n+2\alpha} \int_{B_R(x_0)} |Du-(Du)_{x_0,R}|^2\,dx + \right.$$

(4.18)

$$\left. + \int_{B_R(x_0)} |D(u-v)|^2\,dx\right\} .$$

The last integral can be estimated as in Theorem 4.1, getting

$$\int_{B_R} |D(u-v)|^2\,dx \le c_{18} \int \omega(R^2 + |u-u_{x_0,R}|^2 + |u-v|^2)|Du|^2\,dx$$

and the difference with respect to the scalar case is that we do not know that u is Hölder-continuous.

Now we use the L^q-estimate for Du; more precisely from Section 3, Chapter V, we know that, under our assumptions, there exists a number $q > 2$ such that

$$\left(\fint_{B_R} |Du|^q\,dx\right)^{1/q} \le c_{19}\left(\fint_{B_{2R}} |Du|^2\,dx\right)^{1/2}$$

and we can suppose that q is so close to 2 that $\frac{q\sigma}{q-2} \ge 1$. Then we have

$$\int_{B_R} |D(u-v)|^2\,dx \le c_{20}\left(\int_{B_R} |Du|^q\,dx\right)^{2/q}\left(\int_{B_R} \omega^{\frac{q}{q-2}}\,dx\right)^{1-\frac{2}{q}} \le$$

$$\le c_{21} \int_{B_{2R}} |Du|^2\,dx \left(\fint_{B_R} \omega^{\frac{q}{q-2}}\,dx\right)^{1-\frac{2}{q}} .$$

Since ω is bounded we have

$$\int_{B_R} |D(u-v)|^2 dx \le c_{22} \int_{B_{2R}} |Du|^2 dx . \tag{4.19}$$

Moreover, recalling that $\frac{q}{q-2} > \frac{1}{\sigma}$ and $\omega(t) \le At^\sigma$ we get

$$\omega^{\frac{q}{q-2}} \le c_{23}(R^2 + |u-u_{x_0,R}|^2 + |u-v|^2)$$

and therefore using Poincaré's inequality and (4.19)

$$\fint_{B_R} \omega^{\frac{q}{q-2}} \le c_{24} R^2 \fint_{B_{2R}} |Du|^2 .$$

In conclusion we get

$$\int_{B_R} |D(u-v)|^2 dx \le C_{25} R^{n+2(1-2/q)} \Big(\fint_{B_{2R}} |Du|^2 dx \Big)^{2-2/q}$$

and therefore from (4.18), writing R instead of $2R$:

$$\Phi(x_0,\tau R) \le c_{26} [\tau^{2\alpha} \Phi(x_0,R) + R^{2\left(1-\frac{2}{q}\right)} \tau^{-n} M(x_0,R)^{4\left(1-\frac{1}{q}\right)}]$$

where $0 < \tau < \frac{1}{2}$ and

$$M(x_0,R) = 1 + |(Du)_{x_0,R}| + \Phi(x_0,R)^{1/2} .$$

The proof now proceeds exactly as in the proof of Theorem 2.1, Chapter VI.

q.e.d.

In general $\Omega - \Omega_0$ is nonempty. However in some particular cases one can get regularity everywhere.

The vector valued case: everywhere regularity. Let us consider functionals of the type

$$J[u;\Omega] = \int_\Omega A_{ij}^{\alpha\beta}(x) D_\alpha u^i D_\beta u^j \, dx + \int_\Omega g(x,u)\, dx \tag{4.20}$$

where the coefficients $A_{ij}^{\alpha\beta}$ are Hölder-continuous in Ω with exponent σ and satisfy the Legendre-Hadamard condition

$$A_{ij}^{\alpha\beta}(x)\xi_\alpha \xi_\beta \eta_i \eta_j \geq |\xi|^2 |\eta|^2 \qquad \forall \xi \in R^n, \ \ \forall \eta \in R^N$$

and g is Hölder-continuous with exponent 2σ.

Then, repeating the arguments above, one can easily prove the following (see [116])

THEOREM 4.4. *Let* u *be a minimum for the functional in (4.20). Then* u *has first derivatives Hölder-continuous with any exponent less than* σ *in* Ω.

We do not know whether this result is optimal; however the Hölder exponent of Du cannot reach 2σ, even if $N = 1$ and the functional has the special form

$$\int_\Omega \left(\frac{1}{2} |Du|^2 + |u|^{2\sigma} \right) dx$$

compare with [254].

Finally we want to mention that some of the results of this section can be carried on in order to handle polynomial growth conditions. But the extension of the (partial) C^1-Hölder continuity to general functionals with polynomial growth seems not straightforward.

REFERENCES[1)]

[1] Acerbi, E., N. Fusco – Semicontinuity problems in the calculus of variations, Arch. Rat. Mech. Anal., to appear.

[2] Agmon, S. – Lectures on elliptic boundary value problems, N. J. Van Nostrand, Princeton 1965.

[3] Agmon, S., A. Douglis, L. Nirenberg – Estimates near the boundary for solutions of elliptic partial differential equations satisfying general boundary conditions I, II, Comm. Pure Appl. Math. *12* (1959), 623-727, *17* (1964), 35-92.

[4] Almgren, F. J., Jr. – Existence and regularity almost everywhere of solutions to elliptic variational problems among surfaces of varying topological type and singularity structure, Annals of Math. *87* (1968), 321-391.

[5] ________ – Existence and regularity almost everywhere of solutions to elliptic variational problems with constraints, Mem. Amer. Math. Soc. *165* (1976).

[6] Anzellotti, G. – Dirichlet problem and removable singularities for functionals with linear growth, Boll. UMI Anal. Funz. e Appl. *18* C(1981), 141-159.

[7] ________ – $(1+\omega(R))$-minima, to appear.

[8] Anzellotti, G., M. Giaquinta – Existence of the displacements field for an elasto-plastic body subject to Hencky's law and von Mises yield condition, manuscripta math. *32* (1980), 101-136.

[9] ________ – On the existence of the fields of stresses and displacements for an elasto-perfectly plastic body in static equilibrium, Journ. de Math. pures et appl. *61* (1982), 219-244.

[10] Anzellotti, G., M. Giaquinta, U. Massari, G. Modica, L. Pepe – Note sul problema di Plateau, Ist. Mat. Univ. Pisa, ed. Tecnico Scientifica, Pisa (1974).

[11] Arakcheev, S. A. – Smoothness of generalized solutions of a class of quasi-linear elliptic equations, Vestnik Moskovskogo Universiteta, Matematika *30* (1) (1975), 49-57.

---

1) Without pretending to be exhaustive, this bibliography includes also works not quoted but related to the ones quoted in the text, with the aim of being helpful to the reader.

[12] Aronszajn, M., F. Mulla, P. Szeptycki – On spaces of potential connected with L^p classes, Ann. Inst. Fourier *13*, 2 (1963), 211-306.

[13] Aronszajn, M., K. T. Smith – Theory of Bessel potential, I. Ann. Inst. Fourier *11*, 1 (1961), 385-475.

[14] Aronszajn, M., K. T. Smith, R. Adams – Theory of Bessel potential, II. Ann. Inst. Fourier *17*, 2 (1967), 1-135.

[15] Arzela', C. – Il principio di Dirichlet, Rend. Regia Accad. Bologna (1897).

[16] Attouch, H., C. Sbordone – Asymptotic limits for perturbed functionals of calculus of variations, Ricerche di Mat. *29* (1981), 85-124.

[17] Bakel'man, I. Ya. – Mean curvature and quasilinear elliptic equations, Sibirskii Mat. Ž. *9* (1968). English transl.: Siberian Math. J. *9* (1968), 752-771.

[18] ________ – Geometric problems in quasilinear elliptic equations, Uspehi Mat. Nauk *25* (3) (1970), 49-112. English transl. Russian Math. Surveys *25* (3) (1970), 45-109.

[19] Ball, J. M. – Convexity conditions and existence theorems in nonlinear elasticity, Arch. Rat. Mech. Anal. *63* (1977), 337-403.

[20] ________ – Constitutive inequalities and existence theorems in nonlinear elastostatics, in "Nonlinear Analysis and Mechanics: Heriot-Watt Symposium vol. I" ed. R. J. Knops, Pitman, London 1977.

[21] ________ – On the calculus of variations and sequentially weakly continuous maps, in "Ordinary and partial differential equations" Dundee 1976, Springer Lecture Notes in Math. 564, 13-25.

[22] ________ – Strict convexity, strong ellipticity, and regularity in the calculus of variations, Math. Proc. Camb. Phil. Soc. *87* (1980), 501-513.

[23] Ball, J. M., J. C. Currie, P. J. Olver – Null Lagrangians, weak continuity, and variational problems of arbitrary order, J. Funct. Anal. *41* (1981), 135-174.

[24] Bensoussan, A., J. Frehse – Jeux différentiels stochastiques et systèmes d'équations aux dérivees partielles nonlinéaires, Compt. Rend. Acad. Sc. Paris, to appear.

[25] Berkowitz, L. D. – Lower semicontinuity of integral functionals, Trans. Amer. Math. Soc. *192* (1974), 51-57.

[26] Bernstein, S. – Sur la nature analytique des solutions des équations aux dérivées partielles du second ordre, Math. Ann. *59* (1904), 20-76.

[27] ________ – Sur les équations du calcul des variations, Ann. Sci. Ecole Norm. Sup. *29* (1912), 431-485.

[28] Bers, L., M. Schechter – Elliptic equations, in: "Partial Differential Equations," pp. 131-299, Interscience, New York 1964.

[29] Boccardo, L. – An L^s-estimate for the gradient of solutions of some strongly nonlinear unilateral problems, preprint.

[30] Bombieri, E. – Variational problems and elliptic equations, in: "Proceedings of the International Congress of Mathematicians," Vancouver 1974, vol. I, 53-63.

[31] ———— – Variational problems and elliptic equations (Hilbert's problem 20), in: "Mathematical developments arising from Hilbert problems, Ed. Browder, Proc. Symp. Pure Math. XXVIII (1976).

[32] ———— – Regularity theory for almost minimal currents, Arch. Rat. Mech. and Anal. *78* (1982), 99-130.

[33] Boyarskii, B. V. – Homeomorphic solutions of Beltrami systems, Dokl. Akad. Nauk SSSR *102* (1955), 661-664.

[34] ———— – Generalized solutions of a system of differential equations of the first order of elliptic type with discontinuous coefficients, Mat. Sbornik *43* (1957), 451-503.

[35] Browder, F. E. – Remarks on the direct methods of the calculus of variations, Arch. Rat. Mech. Anal. *20* (1965), 251-258.

[36] ———— – Existence theorems for nonlinear partial differential equations, Proc. Symp. Pure Math. XVI (1970), 1-60.

[37] Caccioppoli, R. – Sulle equazioni ellittiche non lineari a derivate parziali, Rend. Accad. Naz. Lincei *18* (1933), 103-106.

[38] ———— – Sulle equazioni ellittiche a derivate parziali con n variabili indipendenti, Rend. Accad. Naz. Lincei *19* (1934), 83-89.

[39] ———— – Sulle equazioni ellittiche a derivate parziali con due variabili indipendenti e sui problemi regolari del calcolo delle variazioni, Rend. Accad. Naz. Lincei *22* (1935), 305-310, 376-379.

[40] ———— – Limitazioni integrali per le soluzioni di un'equazione lineare ellittica a derivate parziali, Giornale di Mat. di Battaglini *80* (1950-51), 186-212.

[41] Caffarelli, L., R. Kohn, L. Nirenberg – Partial regularity of suitable weak solutions of Navier-Stokes equations, Comm. Pure Appl. Math. 35 (1982), 771-831.

[42] Calderon, A. P., A. Zygmund – On the existence of certain singular integrals, Acta Math. *88* (1952), 85-139.

[43] Campanato, S. – Proprietà di Hölderianità di alcune classi di funzioni, Ann. Sc. Norm. Sup. Pisa *17* (1963), 175-188.

[44] ———— – Proprietà di una famiglia di spazi funzionali, Ann. Sc. Norm. Sup. Pisa *18* (1964), 137-160.

[45] ———— – Equazioni ellittiche del secondo ordine e spazi $\mathcal{L}^{2,\lambda}$, Ann. Mat. Pura e Appl. *69* (1965), 321-380.

[46] ———— – Appunti del corso di analisi superiore, Pisa 1965-66.

[47] Campanato, S. – Su un teorema di interpolazione di G. Stampacchia, Ann. Sc. Norm. Sup., Pisa *20*(1966), 649-652.

[48] ________ – Alcune osservazioni relative alle soluzioni di equazioni ellittiche di ordine 2m, Atti Convegno su Equaz. Deriv. Parz. Bologna (1967), 17-25.

[49] ________ – Maggiorazioni interpolatorie negli spazi $H_{\lambda}^{m,p}(\Omega)$, Ann. Mat. Pura Appl. *75*(1967), 261-276.

[50] ________ – Un risultato relativo ad equazioni ellittiche del secondo ordine di tipo non variazionale, Ann. Sc. Norm. Sup., Pisa *21*(1967), 701-707.

[51] ________ – Equazioni ellittiche non variazionali a coefficienti continui, Ann. Mat. Pura e Appl. *86*(1970), 125-154.

[52] ________ – Partial Hölder continuity of the gradient of solutions of some nonlinear elliptic systems, Rend. Sem. Mat. Padova, *59*(1978), 147-165.

[53] ________ – Sistemi ellittici in forma divergenza. Regolarità all'interno, Quaderni Sc. Norm. Sup., Pisa (1980).

[54] ________ – L^p-regularity and partial Hölder continuity for solutions of second order parabolic systems with strictly controlled growth, Ann. Mat. Pura e Appl. *78*(1980), 287-316.

[55] ________ – L^p-regularity for weak solutions of parabolic systems, Ann. Sc. Norm. Sup. Pisa *7*(1980), 65-85.

[56] ________ – Partial Hölder continuity of solutions of quasilinear parabolic systems of second order with linear growth, Rend. Sem. Mat. Padova *64*(1981), 59-75.

[57] Campanato, S., P. Cannarsa – Differentiability and partial Hölder continuity of the solutions of nonlinear elliptic systems of order 2m with quadratic growth, Ann. Sc. Norm. Sup. Pisa *8*(1981), 287-309.

[58] Campanato, S., G. Stampacchia – Sulle maggiorazioni in L^p nella teoria delle equazioni ellittiche, Boll. UMI *20*(1965), 393-399.

[59] Canfora, A. – Teorema del massimo modulo e teorema di esistenza per il problema di Dirichlet relativo ai sistemi fortemente ellittici, Ricerche di Mat. *15*(1966), 249-294.

[60] Cannarsa, P. – On a maximum principle for elliptic systems with constant coefficients, Rend. Sem. Mat. Padova *64*(1981), 77-84.

[61] Cesari, L. – Lower semicontinuity and lower closure theorems without seminormality condition, Ann. Mat. Pura e Appl. *98*(1974), 381-397.

[62] Coiffman, R.R., C. Fefferman – Weighted norm inequalities for maximal functions and singular integrals, Studia Math. *51*(1974), 241-250.

[63] Colombini, F. – Un teorema di regolarità alla frontiera per soluzioni di sistemi ellittici quasi lineari, Ann. Sc. Norm. Sup. Pisa *25*(1971), 115-161.

[64] Colombini, F., E. De Giorgi, F. Piccinini – Frontiere orientate di misura minima e questioni collegate, Quaderni Sc. Norm. Sup. Pisa (1972).

[65] Courant, R. – Dirichlet's principle, conformal mapping, and minimal surfaces, Interscience, New York (1950).

[66] Courant, R., D. Hilbert – Methods of Mathematical Physics, Interscience, New York (1962).

[67] Dal Maso, G. – Integral representation on $BV(\Omega)$ of Γ-limits of variational integrals, manuscripta math. *30*(1980), 384-416.

[68] Daniljuk, G. N., I. V. Skrypnik – On partial regularity of generalized solutions of quasilinear parabolic systems, Soviet Math. Dokl. *21* (1980), 195-198.

[69] De Giorgi, E. – Sulla differenziabilità e l'analiticità delle estremali degli integrali multipli regolari, Mem. Accad. Sci. Torino cl. Sci. Fis. Mat. Nat. (3) *3*(1957), 25-43.

[70] ———— – Frontiere orientate di misura minima, Quaderni Sc. Norm. Sup. Pisa (1960-61).

[71] ———— – Un esempio di estremali discontinue per un problema variazionale di tipo ellittico, Boll. UMI *4*(1968), 135-137.

[72] ———— – Teoremi di semicontinuità nel calcolo delle variazioni, lezioni tenute all'Ist. Naz. Alta Mat. Roma (1968-69).

[73] Douglis, A., L. Nirenberg – Interior estimates for elliptic systems of partial differential equations, Comm. Pure Appl. Math. *8*(1955), 503-538.

[74] Dubinskii, Yu. A. – Nonlinear elliptic and parabolic equations, Itogi Nauki i Tekhniki, Sovremennye Problemy Matematiki, *9*(1976). Engl. Transl. J. of Soviet Math. *12* (5)(1979), 475-554.

[75] Eells, J. – Regularity of certain harmonic maps, Proc. Durham Symp. on Global Riem. Geo., July 1982, to appear.

[76] Eells, J., L. Lemaire – A report on harmonic maps, Bull. London Math. Soc. *10*(1978), 1-68.

[77] ———— – Selected topics in harmonic maps, N.S.F. Conf. Board Math. Sci., to appear.

[78] Eells, J., J. H. Sampson – Harmonic mappings of Riemannian manifolds, Amer. J. Math. *86*(1964), 109-160.

[79] Eisen, G. – A counterexample for some lower semicontinuity results, Math. Z. *162*(1978), 241-243.

[80] ———— – A selection lemma for sequences of measurable sets, and lower semicontinuity of multiple integrals, manuscripta math. *27*(1979), 73-79.

[81] Ekeland, I. – Nonconvex minimization problems, Bull. Amer. Math. Soc. *1* (1979), 443-474.

[82] Ekeland, I., R. Temam – Analyse convexe et problèmes variationnels, Dunod, Paris (1974).

[83] Evans, L. C. – A new proof of local $C^{1,\alpha}$ regularity for solutions of certain degenerate elliptic P.D.E., preprint (1981).

[84] Federer, H. – Some properties of distributions whose partial derivatives are representable by integration, Bull. Amer. Math. Soc. *74* (1968), 183-186.

[85] ________ – Geometric measure theory, Springer Verlag, Heidelberg and New York (1969).

[86] ________ The singular set of area minimizing rectifiable currents with codimension one and area minimizing flat chains modulo two with arbitrary codimension, Bull. Amer. Math. Soc. *76* (1970), 767-771.

[87] Federer, H., W. P. Ziemer – The Lebesgue set of a function whose distribution derivatives are p-th power summable, Indiana Univ. Math. J. *22* (2) (1972), 139-158.

[88] Fefferman, C., E. M. Stein – H^p spaces of several variables, Acta Math. *129* (1972), 137-193.

[89] Fichera, G. – Problemi elastostatici con vincoli unilaterali: il problema di Signorini con ambigue condizioni al contorno, Mem. Accad. Naz. Lincei *7* (1964), 91-140.

[90] Frehse, J. – On the boundedness of weak solutions of higher order nonlinear elliptic partial differential equations, Boll. UMI *4* (1970), 607-627.

[91] ________ – Una generalizzazione di un controesempio di De Giorgi nella teoria delle equazioni ellittiche, Boll. UMI *6* (1970), 998-1002.

[92] ________ – A discontinuous solution of a mildly nonlinear elliptic system, Math. Z. *134* (1973), 229-230.

[93] ________ – A note on the Hölder continuity of solutions of variational problems, Abhandlung. Math. Sem. Hamburg *43* (1975), 59-63.

[94] ________ – Essential self-adjointness of singular elliptic operators, Boll. Soc. Bras. Mat. *8* (1977), 87-107.

[95] ________ – On two-dimensional quasi-linear elliptic systems, manuscripta math. *28* (1979), 21-50.

[96] ________ – Un problème variationnel bidimensionel possédant des extremales bornées et discontinues, C. R. Acad. Sc. Paris *289* sér. A. (1979), 751-753.

[97] Frehse, J., U. Mosco – Variational inequalities with one-sided irregular obstacle, manuscripta math. *28* (1979), 219-233.

[98] Frehse, J., U. Mosco – Irregular obstacles and quasi-variational inequalities of stochastic impulse control, Ann. Sc. Norm. Sup. Pisa *9*(1982), 105-157.

[99] Friedrichs, K. O. – On the differentiability of the solutions of linear elliptic differential equations, Comm. Pure Appl. Math. *6*(1953), 299-326.

[100] Fubini, G. – Sul principio di Dirichlet, Rend. Circ. Mat. Palermo *22*(1906), 383-386.

[101] Fusco, N. – Quasi-convessità e semicontinuità per integrali multipli di ordine superiore, Ricerche di Mat. *29*(1980), 307-323.

[102] Gårding, L. – Dirichlet problem for linear elliptic partial differential equations, Math. Scand. *1*(1953), 55-72.

[103] Gehring, F. W. – The L^p-integrability of the partial derivatives of a quasi conformal mapping, Acta Math. *130*(1973), 265-277.

[104] Giaquinta, M. – Sistemi ellittici non lineari, Convegno su: Sistemi ellittici non lineari ed applicazioni. Ferrara 20-24 Settembre (1977).

[105] ———— – A counterexample to the boundary regularity of solutions to elliptic quasilinear systems, manuscripta math. *14*(1978), 217-220.

[106] ———— – Sistemi ellittici nonlineari. Teoria della regolarità, Boll. UMI (5) *16-A*(1979), 259-283.

[107] ———— – A note on the higher order integrability from reverse Hölder inequality, Ist. Mat. Appl. Firenze (1980).

[108] ———— – Regularity results for weak solutions to variational equations and inequalities for nonlinear elliptic systems, Preprint No. 54 SFB 123, Heidelberg (1980).

[109] ———— – Remarks on the regularity of weak solutions to some variational inequalities, Math. Z. *177*(1981), 15-31.

[110] ———— – On the differentiability of the extremals of variational integrals, Proc. Spring School on "Nonlinear Analysis, Function spaces and Applications 2," Pisek, May 24-28, 1982, Teubner, Leipzig.

[111] ———— – The regularity problem of extremals of variational integrals, Proc. NATO/LMS Advanc. Study Inst. on "Systems of nonlinear partial differential equations" Oxford, July 25 - August 7, to appear.

[112] Giaquinta, M., E. Giusti – Partial regularity for the solution to nonlinear parabolic systems, Ann. Mat. Pura e Appl. *47*(1973), 253-266.

[113] ———— – Nonlinear elliptic systems with quadratic growth, manuscripta math. *24*(1978), 323-349.

[114] ———— – On the regularity of the minima of variational integrals, Acta Math. *148*(1982), 31-46.

[115] Giaquinta, M., E. Giusti – The singular set of the minima of certain quadratic functionals, preprint 453 SFB72 Bonn (1981), to appear in Ann. Sc. Norm. Sup. Pisa.

[116] ———— – Differentiability of minima of nondifferentiable functionals, Inventiones Math., to appear.

[117] ———— – Q-minima, to appear.

[118] Giaquinta, M., S. Hildebrandt – Estimation a priori des solutions faibles de certains systèmes nonlineaires elliptiques, Sem. Goulaouic-Meyer-Schwartz 1980-81, exposé 17, Mars 1981.

[119] ———— – A priori estimates for harmonic mappings, J. reine u. angew. Math. *336*(1982), 124-164.

[120] Giaquinta, M., O. John, J. Stara – The example showing the exactness of the number conditioning regularity of weak solutions of quasilinear elliptic systems of 2nd order, preprint.

[121] Giaquinta, M., G. Modica – Regularity results for some classes of higher order nonlinear elliptic systems, J. für reine u angew. Math. *311/312*(1979), 145-169.

[122] ———— – Almost-everywhere regularity results for solutions of nonlinear elliptic systems, manuscripta math. *28*(1979), 109-158.

[123] ———— – Nonlinear systems of the type of the stationary Navier-Stokes system, J. für reine u. angew. Math. *330*(1982), 173-214.

[124] Giaquinta, M., G. Modica, J. Souček – Functionals with linear growth in the calculus of variations I, II, Comment. Math. Univ. Carolinae *20*(1979), 143-156, 157-172.

[125] Giaquinta, M., J. Nečas – On the regularity of weak solutions to nonlinear elliptic systems via Liouville's type property, Comment. Univ. Carolinae *20*(1979), 111-121.

[126] ———— – On the regularity of weak solutions to nonlinear elliptic systems of partial differential equations, J. für reine u. angew. Math. *316*(1980), 140-159.

[127] Giaquinta, M., M. Struwe – An optimal regularity result for a class of quasilinear parabolic systems, manuscripta math. *36*(1981), 223-239.

[128] ———— – On the partial regularity of weak solutions of nonlinear parabolic systems, Math. Z. *179*(1982), 437-451.

[129] Gilbarg, D., N.S. Trudinger – Elliptic partial differential equations of second order, Springer Verlag, Heidelberg, New York (1977).

[130] Giusti, E. – Precisazione delle funzioni $H^{1,p}$ e singolarità delle soluzioni deboli di sistemi ellittici non lineari, Boll. UMI *2*(1969), 71-76.

[131] Giusti, E. – Regolarità parziale delle soluzioni di sistemi ellittici quasi lineari di ordine arbitrario, Ann. Sc. Norm. Sup. Pisa *23* (1969), 115-141.

[132] ———— – Un'aggiunta alla mia nota: Regolarità parziale delle soluzioni di sistemi ellittici quasi lineari di ordine arbitrario, Ann. Sc. Norm. Sup. Pisa *27* (1973), 161-166.

[133] ———— – Boundary value problems for nonparametric surfaces of prescribed mean curvature, Ann. Sc. Norm. Sup. Pisa *3* (1976), 501-548.

[134] ———— – Minimal surfaces and functions of bounded variation, Note on Pure Math. *10* (1977).

[135] ———— – Equazioni ellittiche del secondo ordine, Quaderni dell'Un. Mat. Italiana *6* (1968) Ed. Pitagora, Bologna.

[136] ———— – Generalized solutions for the mean curvature equation, Pacific J. Math. *88* (1980), 297-321.

[137] ———— – Regularity and singularities of solutions of nonlinear elliptic systems, Proc. NATO/LMS Advanc. Study Inst. on "Systems of nonlinear partial differential equations" Oxford, July 25 - August 7, 1982, to appear.

[138] Giusti, E., M. Miranda – Un esempio di soluzioni discontinue per un problema di minimo relativo ad un integrale regolare del calcolo delle variazioni, Boll. UMI *2* (1968), 1-8.

[139] ———— – Sulla regolarità delle soluzioni deboli di una classe di sistemi ellittici quasilineari, Arch. Rat. Mech. Anal. *31* (1968), 173-184.

[140] Giusti, E., G. Modica – A note on the regular points for solutions of elliptic systems, manuscripta math. *29* (1979), 417-426.

[141] Granlund, S. – An L^p-estimate for the gradient of extremals, Math. Scand. *50* (1982), 66-72.

[142] Grüter, M. – Ein Hebbarkeitssatz für isolierte Singularitäten, Bonn, SFB72 (1978).

[143] ———— – Über die Regularität schwacher Lösungen des Systems $\Delta x = 2H(x) x_u \wedge x_v$, Dissertation, Düsseldorf (1979).

[144] ———— – Regularity of weak H-surfaces, J. für reine u. angew. Math. *329* (1981), 1-15.

[145] Gurov, L. G., Yn. G. Reshetnyak – An analogue of the concept of functions with bounded mean oscillation, Sibirskii Math. Z. *17* (1976), 540-546.

[146] Haar, A. – Über das Plateausche Problem, Math. Ann. *97* (1926), 124-158.

[147] Hadamard, J. – Sur le principle de Dirichlet, Bull. Soc. Math. France *34*(1906), 135-138.

[148] Hamilton, R. S. – Harmonic maps of manifolds with boundary, Springer Lecture Notes 471 (1975).

[149] Hartman, P., G. Stampacchia – On some nonlinear elliptic differential-functional equations, Acta Math. *115*(1966), 271-310.

[150] Heinz, E. – On certain nonlinear elliptic differential equations and univalent mappings, J. Analyse math. *5*(1956/57), 197-272.

[151] ________ – Existence theorems for one-to-one mappings associated with elliptic systems of second order I, II, J. Analyse math. *15*(1965), 325-352; *17*(1966), 145-184.

[152] ________ – Ein regularitätssatz für schwache Lösungen nicht linearer elliptischer Systeme, Nachr. Akad. Wiss. Göttingen 1975 No. 1, 1-13.

[153] Hempel, R. – Eine Variationsmethode für elliptische Differentialoperatoren mit strengen Nichtlinearitäten, J. für reine u angew Math. *333*(1982), 179-190.

[154] Hilbert, D. – Über das Dirichletsche Prinzip, Jber. Deutsch. Math. Verein *8*(1900), 184-188.

[155] ________ – Über das Dirichletsche Prinzip, Math. Ann. *59*(1904), 161-168.

[156] Hildebrandt, S. – Variationsrechnung Mehrdimensionaler Integrale, Teil 2, Univ. Bonn (1975).

[157] ________ – Variationsrechnung und Hamiltonsche Mechanik, Bonn, Sommersemester 1977.

[158] ________ – Regularity results for solutions of quasilinear elliptic systems, Convegno su: Sistemi ellittici nonlineari ed applicazioni, Ferrara 20-24 Settembre 1977.

[159] ________ – Liouville theorems for harmonic mappings and an approach to Bernstein theorem, Annals of Math. Studies 102, Princeton (1982), 107-132.

[160] ________ – Nonlinear elliptic systems and harmonic mappings, Vorlesungsreihe SFB72 No. 3 (1980); Proc. Beijing Symp. Diff. Geo. and Diff. Eq., to appear.

[161] ________ – Elliptic systems of P.D.E., Proc. NATO/LMS Advanc. Study Inst. on "Systems of nonlinear partial differential equations," Oxford, July 25 - August 7, 1982, to appear.

[162] Hildebrandt, S., J. Jost, K. -O. Widman – Harmonic mappings and minimal submanifolds, Inventiones Math. *62*(1980), 269-298.

[163] Hildebrandt, S., H. Kaul, K. -O. Widman – An existence theorem for harmonic mappings of Riemannian manifolds, Acta Math. *138*(1977), 1-16.

[164] Hildebrandt, S., K.-O. Widman – Some regularity results for quasilinear elliptic systems of second order, Math. Z. *142*(1975), 67-86.

[165] ———— – On the Hölder continuity of weak solutions of quasilinear elliptic systems of second order, Ann. Sc. Norm. Sup. Pisa *4*(1977), 145-178.

[166] ———— – Variational inequalities for vector valued functions, J. für reine u. angew. Math. *309*(1979), 181-220.

[167] ———— – Sätze von Liouvilleschen Typ für quasilineare elliptische Gleichungen und Systeme, Nachr. Akad. Wiss. Göttingen II, *4*(1979), 41-59.

[168] Hopf, E. – Zum analytischen Charakter der Lösungen regulärer zweidimensionaler Variationsprobleme, Math. Z. *30*(1929), 404-413.

[169] ———— – Über den funktionalen, insbesondere den analytischen Charakter der Lösungen elliptischer Differentialgleichungen zweiter Ordnung, Math. Z. *34*(1932), 194-233.

[170] Iofre, A.D. – On lower semicontinuity of integral functionals I., SIAM I. Control optimization, *15*(1977), 521-538.

[171] Ivert, P.A. – Regularitätsuntersuchungen von Lösungen elliptischer Systeme von quasilinearen Differentialgleichungen zweiter Ordnung, Dissertation, Linköping (1978).

[172] ———— – Regularitätsuntersuchungen von Lösungen elliptischer Systeme von quasilinearen Differentialgleichungen zweiter Ordnung, manuscripta math. *30*(1979), 53-88.

[173] ———— – On quasilinear elliptic systems in diagonal form, Math. Z. *170*(1980), 283-286.

[174] ———— – Partial regularity of vector valued functions minimizing variational integrals, preprint.

[175] Iwaniec, T. – Projections onto gradient fields and L^p-estimates for degenerated elliptic operators, to appear.

[176] Jäger, W. – Ein Maximumprinzip für ein System nichtlinearer Differentialgleichungen, Nachr. Akad. Wiss. Göttingen *11*(1976).

[177] Jager, W., H. Kaul – Uniqueness and stability of harmonic maps and their Jacobi fields, manuscripta math. *28*(1979), 269-291.

[178] John, F., L. Nirenberg – On functions of bounded mean oscillation, Comm. Pure Appl. Math. *14*(1961), 415-426.

[179] Jost, J. – Ein Existenzbeweis für harmonische Abbildungen, die ein Dirichletproblem lösen, mittels der Methode des Wärmeflusses, manuscripta math. *34*(1981), 17-25.

[180] ———— – Univalency of harmonic mappings between surfaces, J. für reine u. angew. Math. *324*(1981), 141-153.

[181] Jost, J., M. Meier – Boundary regularity for minima of certain quadratic functionals, Math. Ann., to appear.

[182] Jost, J., R. Schoen – On the existence of harmonic diffeomorphisms between surfaces, Inventiones Math. *66*(1982), 353-359.

[183] Kadleč, J., J. Nečas – Sulla regolarità delle soluzioni di equazioni ellittiche negli spazi $H^{k,\lambda}$, Ann. Sc. Norm. Sup. Pisa *21*(1967), 527-545.

[184] Kawohl, B. – On Liouville theorem, continuity and Hölder continuity of weak solutions to some quasilinear elliptic systems, Comment. Math. Univ. Carolinae *21* (1980), 679-697.

[185] Kinderlehrer, D., G. Stampacchia – Variational inequalities and applications, Academic Press, New York (1980).

[186] Koshelev, A. I. – Regularity of solutions of quasilinear elliptic systems, Uspekhi Mat. Nauk *33*(1978), 3-49; Engl. Transl. Russian Math. Surv. *33*(1978), 1-52.

[187] Kufner, A., O. John, S. Fučik – Function spaces, Akademia, Praha (1977).

[188] Ladyzhenskaya, O. A., N. N. Ural'tseva – On the smoothness of weak solutions of quasilinear equations in several variables and of variational problem, Comm. Pure Appl. Math. *14*(1961), 481-495.

[189] ________ – Quasilinear elliptic equations and variational problems with several independent variables, Uspekhi Math. Nauk *16*(1961), 19-90; Engl. Transl. Russian Math. Surv. *16*(1961), 17-92.

[190] ________ – On Hölder continuity of solutions and their derivatives of linear and quasilinear elliptic and parabolic equations, Trudy Mat. Inst. Steklov *73*(1964), 172-220; Engl. Transl. Amer. Math. Soc. Transl. (2) *61* (1967), 207-269.

[191] ________ – Linear and quasilinear elliptic equations, Moscow, Nauka (1964); Engl. Transl. Academic Press New York (1968); Second Russian edition: Nauka (1973).

[192] Landes, R. – Quasilinear elliptic operators and weak solutions of the Euler equations, manuscripta math. *27*(1979), 47-72.

[193] Lawson, H. B., R. Osserman – Non-existence, non-uniqueness and irregularity of solutions to the minimal surface system, Acta Math. *139*(1977), 1-17.

[194] Lebesgue, H. – Sur le problème de Dirichlet, Rend. Circ. Mat. Palermo *24*(1907), 371-402.

[195] Lemaire, L. – Applications harmoniques de surfaces riemanniennes, J. Diff. Geo. *13*(1978), 61-88.

[196] ________ – Boundary value problems for harmonic and minimal maps of surfaces into manifolds, Ann. Sc. Norm. Sup. Pisa *9*(1982), 91-103.

[197] Leray, J. – Etude de diverses équations intégrales nonlineaires et de quelques problèmes qui pose l'hydrodynamique, Journ. Math. Pures et Appl. *12* (1933), 1-82.

[198] ——— – Majoration des dérivées secondes des solutions d'un problème de Dirichlet, Journ. de Math. Pure et Appl. *17* (1938), 89-104.

[199] Leray, J., J. L. Lions – Quelques résultats de Višik sur les problèmes elliptiques nonlinéaires par la méthode de Minty-Browder, Bull. Soc. Math. France *93* (1965), 97-107.

[200] Leray, H., H. Schauder – Topologie et équations fonctionnelles, Ann. Sci. Ecol. Norm. Sup. *51* (1934), 45-78.

[201] Levi, B. – Sul principio di Dirichlet, Rend Circ. Mat. Palermo *22* (1906), 293-360.

[202] Liechtenstein, L. – Über den analytischen Charakter der Lösungen zweidimensionaler Variationsprobleme, Bull. Acad. Sci. Cracoviae (1912).

[203] ——— – Neuere Entwicklung der Theorie partieller Differentialgleichungen zweiter Ordnung von elliptischen Typus, in: Enc. d. Math. Wissensch. 2.3.2 (1924), 1277-1334. Teubner, Leipzig 1923-1927.

[204] Lions, J. L. – Quelques méthodes de résolution des problèmes aux limites nonlinéaires, Dunod, Paris (1969).

[205] Luckhaus, S. – Existence and regularity of weak solutions to the Dirichlet problem for semilinear elliptic systems of higher order, J. für reine u. angew. Math. *306* (1979), 192-207.

[206] Marcellini, P., C. Sbordone – Semicontinuity problems in the calculus of variations, Nonlinear Anal. Theory, Meth. and Appl. *4* (1980), 241-257.

[207] ——— – On the existence of minima of multiple integrals of the calculus of variations, Preprint (1981).

[208] Maz'ya, V. G. – Examples of nonregular solutions of quasilinear elliptic equations with analytic coefficients, Funktsional'nyi Analiz. i Ego Prilosheniya *2* (1968), 53-57.

[209] Meier, M. – Liouville theorems for nonlinear elliptic equations and systems, manuscripta math. *29* (1979), 207-228.

[210] ——— – Liouville theorems for nondiagonal elliptic systems in arbitrary dimensions, Math. Z. *176* (1981), 123-133.

[211] ——— – Boundedness and integrability properties of weak solutions of quasilinear elliptic systems, J. für reine u. angew. Math. *333* (1982), 191-220.

[212] Meier, M. – Liouville theorems, partial regularity, and Hölder continuity of weak solutions to quasilinear elliptic systems, Preprint.

[213] Meyers, N. G. – An L^p-estimate for the gradient of solutions of second order elliptic divergence equations, Ann. Sc. Norm. Sup. Pisa (3) *17*(1963), 189-206.

[214] ________ – Mean oscillation over cubes and Hölder continuity, Proc. Am. Math. Soc. *15*(1964), 717-721.

[215] ________ – Quasi-convexity and lower semicontinuity of multiple variational integrals of any order, Trans. Amer. Math. Soc. *119*(1965), 125-149.

[216] Meyers, N. G., A. Elcrat – Some results on regularity for solutions of nonlinear elliptic systems and quasiregular functions, Duke Math. J. *42*(1975), 121-136.

[217] Miranda, C. – Sul problema misto per le equazioni lineari ellittiche, Ann. Mat. Pura e Appl. *39*(1955), 279-303.

[218] ________ – Partial differential equations of elliptic type, 2nd ed. Springer Verlag, Heidelberg, New York (1970).

[219] Miranda, M. – Existence and regularity of hypersurfaces of R^n with prescribed mean curvature, Proc. Symp. Pure Math. *23*(1973).

[220] Morrey, C. B. Jr. – On the solutions of quasi-linear elliptic partial differential equations, Trans. Amer. Math. Soc. *43*(1938), 126-166.

[221] ________ – Existence and differentiability theorems for the solutions of variational problems for multiple integrals, Bull. Amer. Math. Soc. *46*(1940), 439-458.

[222] ________ – Multiple integral problems in the calculus of variations and related topics, Univ. California Publ. Math. *1*(1943), 1-130.

[223] ________ – The problem of Plateau on a Riemannian manifold, Annals of Math. *49*(1948), 807-851.

[224] ________ – Quasi-convexity and the lower semicontinuity of multiple integrals, Pacific J. Math. *2*(1952), 25-53.

[225] ________ – Second order elliptic systems of differential equations, Ann. of Math. Studies No. 33, Princeton Univ. Press (1954), 101-159.

[226] ________ – On the analyticity of the solutions of analytic nonlinear elliptic systems of partial differential equations I, II, Amer. J. of Math. *80*(1958), 198-218, 219-234.

[227] ________ – Second order elliptic equations in several variables and Hölder continuity, Math. Z. *72*(1959), 146-164.

[228] ________ – Multiple integral problems in the calculus of variations and related topics, Ann. Sc. Norm. Sup. Pisa *14*(1960), 1-61.

[229] Morrey, C.B. Jr. – Existence and differentiability theorems for variational problems for multiple integrals, Part. Diff. Eq. and Cont. Mech. Univ. Wisconsin Press, Madison (1961), 241-270.

[230] ——— – Some recent developments in the theory of partial differential equations, Bull. Amer. Math. Soc. *68* (1962), 279-297.

[231] ——— – Multiple integrals in the calculus of variations, Springer Verlag, Heidelberg, New York, (1966).

[232] ——— – Partial regularity results for nonlinear elliptic systems, Journ. Math. and Mech. *17* (1968), 649-670.

[233] ——— – Differentiability theorems for nonlinear elliptic equations, Actes Congrès Intern. Math. (1970).

[234] Moser, J. – A new proof of De Giorgi's theorem concerning the regularity problem for elliptic differential equations, Comm. Pure Appl. Math. *13* (1960), 457-468.

[235] ——— – On Harnack's theorem for elliptic differential equations, Comm. Pure Appl. Math. *14* (1961), 577-591.

[236] Murat, F. – Compacité par compensation II, Proc. Int. Meeting on Recent Methods in Nonlinear Analysis, Roma (1978), Edited by E. De Giorgi, E. Magenes, U. Mosco.

[237] Nash, J. – Continuity of solutions of parabolic and elliptic equations, Amer. J. Math. *8* (1958), 931-954.

[238] Nečas, J. – Sur la régularité des solutions variationnelles des équations elliptiques non-linéaires d'ordre 2k en deux dimensions, Ann. Sc. Norm. Sup. Pisa *21* (1967), 427-457.

[239] ——— – Les Méthodes directes en théorie des équations elliptiques, Praha, Akademia (1967).

[240] ——— – Sur la régularité des solutions faibles des équations elliptiques nonlinéaires, Comment. Math. Univ. Carolinae *9, 3* (1968), 365-413.

[241] ——— – Example of an irregular solution to a nonlinear elliptic system with analytic coefficients and conditions for regularity, in: Theory of Non Linear Operators, Abhandlungen Akad. der Wissen. der DDR (1977), Proc. of a Summer School held in Berlin (1975).

[242] ——— – On the regularity of weak solutions to variational equations and inequalities for nonlinear second order elliptic systems, Equadiff IV, Praha, Springer Verlag Lectures Notes No. 703.

[243] Nečas, J., O. John, J. Stará – Counterexample to the regularity of weak solutions of elliptic systems, Comment. Math. Univ. Carolinae *21* (1980), 145-154.

[244] Nečas, J., J. Stará – Principio di massimo per i sistemi ellittici quasilineari nondiagonali, Boll. UMI *6*(1972), 1-10.

[245] Nečas, J., J. Stará, R. Švarc – Classical solution to a second order nonlinear elliptic system in R_3, Ann. Sc. Norm. Sup. Pisa *5*(1978), 605-631.

[246] Nečas, J., M. Stipl – A paradox in the theory of linear elasticity, Aplikace Mat. *21*(1976), 431-433.

[247] Nirenberg, L. – Remarks on strongly elliptic partial differential equations, Comm. Pure Appl. Math. *8*(1955), 649-675.

[248] ———— – On elliptic partial differential equations, Ann. Sc. Norm. Sup. Pisa *13*(1959), 115-162.

[249] Nitsche, J.C.C. – Vorlesungen über Minimalflächen, Springer Verlag, Heidelberg, New York (1975).

[250] Olech, C. – A characterization of L^1-weak lower semicontinuity of integral functionals, Bull. Acad. Polon. Sci. *25*(1977), 135-142.

[251] Peetre, J. – Spaces $L^{p,\lambda}$ and interpolation, J. Functional Analysis *4*(1969), 71-87.

[252] Pepe, L. – Risultati di regolarità parziale per le soluzioni $H^{1,p}(\Omega), 1 < p < 2$, di sistemi ellittici quasilineari, Ann. Univ. Ferrara *8*(1971), 129-148.

[253] Petrowski, I. – Sur l'analyticité des solutions des systèmes d'équations différentielles, Rec. Math. N.S. Mat. Sbornik *5*(1939), 3-70.

[254] Phillips, D. – A minimization problem and the regularity of solutions in the presence of a free boundary, Preprint.

[255] Piccinini, L.C. – Proprietà di inclusione e interpolazione tra spazi di Morrey e loro generalizzazioni, Publ. Sc. Norm. Sup. Pisa (1969).

[256] Rado, T. – On the problem of Plateau, Springer, Berlin (1933).

[257] Reshetnyak, Yu. G. – Stability estimates in Liouville's theorem and the L^p-integrability of the derivatives of quasi-conformal mappings, Sibirskii Math. J. *17*(1976), 868-896.

[258] Sacks, J., K. Uhlenbeck – The existence of minimal immersions of 2-spheres, Annals of Math. *113*(1981), 1-24.

[259] Saks, S. – Theory of the integral, New York (1938).

[260] Schauder, J. – Über lineare elliptische Differentialgleichungen zweiter Ordnung, Math. Z. *38*(1934), 257-282.

[261] Schoen, R., K. Uhlenbeck – A regularity theory for harmonic maps, J. Diff. Geo. *17*(1982), 307-335.

[262] Schoen, R., K. Uhlenbeck – Boundary regularity and miscellaneous results on harmonic maps, J. Diff. Geo., to appear.

[263] Schulze B. W., G. W. Wildenhein – Methoden der Potentialtheorie für elliptische Differentialgleichungen beliebiger Ordnung, Akademic-Verlag, Berlin (1977).

[264] Serrin, J. – On a fundamental theorem of the calculus of variations, Acta Math. *102* (1959), 1-32.

[265] ——— – On the definition and properties of certain variational integrals, Trans. Amer. Math. Soc. *101* (1961), 139-167.

[266] ——— – Pathological solutions of elliptic equations, Ann. Sc. Norm. Sup. Pisa *18* (1964), 365-387.

[267] ——— – Local behavior of solutions of quasilinear elliptic equations, Acta Math. *111* (1964), 247-302.

[268] ——— – The problem of Dirichlet for quasilinear elliptic differential equations with many independent variables, Philos. Trans. Roy. Soc. London ser. A *264* (1969), 413-496.

[269] ——— – The solvability of boundary value problems (Hilbert problem 19), in: Mathematical developments arising from Hilbert problems, Ed. Browder, Proc. Symp. Pure Math. XXVIII (1976).

[270] Skrypnik, I. V. – On the regularity of generalized solutions of quasilinear elliptic equations of arbitrary order, Dokl. Akad. Nauk SSSR *203* No. 1 (1972), 36-38.

[271] ——— – On the continuity of generalized solutions of elliptic equations of higher order, Doporidi Akad. Nauk Ukr. SSSR, su. A, No. 1 (1972), 43-45.

[272] ——— – Nonlinear elliptic equations of higher order, Ed. Naukova Dumka Kiev (1973).

[273] ——— – Solvability and properties of solutions of nonlinear elliptic equations, Itogi Nauki i Tekhniki, Sovremennye Problemy Mat. *9* (1976); Engl. Transl. J. Soviet Math. *12*, 5 (1979), 555-620.

[274] Souček, J. – private communication, April 1980.

[275] Sperner, E., Jr. – Ein Regularitätssastz für Systeme nichtlinearer elliptischer Differentialgleichungen, Math. Z. *146* (1976), 57-68.

[276] ——— – A priori gradient estimates for harmonic mappings, preprint.

[277] Stampacchia, G. – Problemi al contorno ellittici, con dati discontinui, dotati di soluzioni Hölderiane, Ann. Mat. Pura e Appl. *51* (1960), 1-38.

[278] ——— – Sur les espaces de functions qui interviennent dans les problèmes aux limites elliptiques, Coll. Anal. Fonct. C.B.R.M. Louvain (1960).

[279] Stampacchia, G. – On some regular multiple integral problems in the calculus of variations, Comm. Pure Appl. Math. *16*(1963), 383-421.

[280] ________ – $L^{p,\lambda}$ spaces and interpolation, Comm. Pure Appl. Math. *17*(1964), 293-306.

[281] ________ – The spaces $L^{p,\lambda}$, $N^{(p,\lambda)}$ and interpolation, Ann. Sc. Norm. Sup. Pisa *19*(1965), 443-462.

[282] ________ – Le problème de Dirichlet pour les équations elliptiques du second ordre a coefficients discontinues, Ann. Inst. Fourier *15*, 1,(1965), 189-258.

[283] ________ – Equations elliptiques du second ordre à coefficients discontinues, Les Presses de l'Univ. de Montréal (1966).

[284] Stará, J. – Regularity results for nonlinear elliptic systems in two dimensions, Ann. Sc. Norm. Sup. Pisa *25*(1971), 163-190.

[285] Stein, E. M. – Singular integrals and differentiability properties of functions, Princeton Univ. Press, Princeton (1970).

[286] Stein, E. M., G. Weiss – Introduction to Fourier Analysis on Euclidean spaces, Princeton University Press, Princeton (1971).

[287] Stredulinsky, E. W. – Higher integrability from reverse Hölder inequalities, Indiana Univ. Math. I. *29*, 3, (1980), 408-417.

[288] Struwe, M. – A counterexample in elliptic regularity theory, manuscripta math. *34*(1981), 85-92.

[289] ________ – On the Hölder continuity of bounded weak solutions of quasilinear parabolic systems, manuscripta math. *35*(1981), 125-145.

[290] ________ – A counterexample in regularity theory for parabolic systems, Comment. Univ. Carolinae, to appear.

[291] Tachikawa, A. – On interior regularity and Liouville's type theorem for harmonic mappings, manuscripta math., to appear.

[292] Tausch, E. – A class of variational problems with linear growth, Math. Z. *164*(1975), 159-191.

[293] Todorov, T. G. – Smoothness of generalized solutions of quasilinear elliptic systems of higher order, Problemy Mat. Analize, *5*(1975), 180-191.

[294] Tonelli, L. – Fondamenti di calcolo delle variazioni, 2 vol., Zanichelli, Bologna (1921-23).

[295] Uhlenbeck, K. – Harmonic maps: A direct method in the calculus of variations, Bull. Amer. Math. Soc. *76*(1970), 1082-1087.

[296] ________ – Regularity for a class of nonlinear elliptic systems, Acta Math. *138*(1977), 219-240.

[297] Višik, I. M. – Quasilinear strongly elliptic systems of differential equations in divergence form, Trudy Mosk. Mat. Obšč. *12*(1963), 140-208.

[298] von Wahl, W. – Existenzsätze für nichtlineare elliptische Systeme, Nachr. Akad. Wissen. Göttingen *3*(1978), 53-62.

[299] von Wahl, W., M. Wiegner – Über die Hölderstetigkeit schwacher Lösungen semilinearer elliptischer Systeme mit einseitiger Bedingung, manuscripta math. *19*(1976), 385-408.

[300] Widman, K.-O. – Hölder continuity of solutions of elliptic systems, manuscripta math. *5*(1971), 299-308.

[301] ———— – Local bounds for solutions of higher order nonlinear elliptic partial differential equations, Math. Z. *121*(1971), 81-95.

[302] Wiegner, M. – Ein optimaler Regularitätssatz für schwache Lösungen gewisser elliptischer Systeme, Math. Z. *147*(1976), 21-28.

[303] ———— – A priori Schranken für Lösungen gewisser elliptischer Systeme, manuscripta math. *18*(1976), 279-297.

[304] ———— – Das Existenz- und Regularitätsproblem bei Systemen nichtlinearer elliptischer Differentialgleichungen, Habilitationsschrift, Bochum (1977).

[305] ———— – Regularity theorems for nondiagonal elliptic systems, Arkiv für Math. *20*(1982), 1-13.

[306] ———— – On two dimensional elliptic systems with a one-sided condition, Math. Z. *178*(1981), 493-500.

[307] Zygmund, A. – Trigonometric series, Cambridge University Press (1959).

Library of Congress Cataloging in Publication Data

Giaquinta, Mariano, 1947-
Multiple integrals in the calculus of variations and non linear elliptic systems.

(Annals of mathematics studies ; 105)
Bibliography: p.
1. Calculus of variations. 2. Integrals, Multiple. 3. Differential equations, Elliptic. I. Title. II. Series.
QA315.G47 1983 515'.64 82-24072
ISBN 0-691-08330-4
ISBN 0-691-08331-2 (pbk.)

Mariano Giaquinta is Professor of Mathematics at the University of Florence.